# Photochemistry and Photophysics

# Photochemistry and Photophysics

Editor: Derek Atkinson

STATES
ACADEMIC PRESS
www.statesacademicpress.com

Published by States Academic Press
109 South 5th Street,
Brooklyn, NY 11249, USA
www.statesacademicpress.com

Photochemistry and Photophysics
Edited by Derek Atkinson

© 2022 States Academic Press

International Standard Book Number: 978-1-63989-412-3 (Hardback)

**Cataloging-in-Publication Data**

Photochemistry and photophysics / edited by Derek Atkinson.
      p. cm.
Includes bibliographical references and index.
ISBN 978-1-63989-412-3
1. Photochemistry. 2. Optics. 3. Chemistry, Physical and theoretical.
4. Photoionization. I. Atkinson, Derek.
QD708.2 .P46 2022
541.35--dc23

# Contents

# Preface

Photochemistry is the branch of chemistry concerned with the chemical effects of light. It is generally used to describe a chemical reaction caused by the absorption of ultraviolet radiation, visible light or infrared radiation. A molecule in its ground state can absorb light energy and go to a higher energy state. Photochemistry is of immense importance in nature as it is the basis of photosynthesis, vision and the formation of vitamin D with sunlight. Photophysics is a branch of science that deals with the physical properties of matter affected by light, and the physical effects of light. It mainly studies the processes that occur when sunlight, filtered through the Earth's atmosphere, interacts with matter present on the Earth. This book unfolds the innovative aspects of photochemistry and photophysics, which will be crucial for the progress of this field in the future. It presents researches and studies performed by experts across the globe. This book will prove to be immensely beneficial to students and researchers in this field.

This book is a result of research of several months to collate the most relevant data in the field.

When I was approached with the idea of this book and the proposal to edit it, I was overwhelmed. It gave me an opportunity to reach out to all those who share a common interest with me in this field. I had 3 main parameters for editing this text:

1. Accuracy – The data and information provided in this book should be up-to-date and valuable to the readers.

2. Structure – The data must be presented in a structured format for easy understanding and better grasping of the readers.

3. Universal Approach – This book not only targets students but also experts and innovators in the field, thus my aim was to present topics which are of use to all.

Thus, it took me a couple of months to finish the editing of this book.

I would like to make a special mention of my publisher who considered me worthy of this opportunity and also supported me throughout the editing process. I would also like to thank the editing team at the back-end who extended their help whenever required.

Editor

# Photophysics and Photochemistry of Conformationally Restricted Triarylmethanes: Application as Photoredox Catalysts

Sankalan Mondal and Satyen Saha

**Abstract**

The three aryl rings of triarylmethanes are free to rotate. However this free rotation can be restricted on either by bridging the aryl rings through covalent bonds or through heteroatoms resulting in the formation of Conformationally Restricted Triarylmethanes (CRT). The photophysics and photochemistry of these CRTs, like 9-arylxanthenes (oxygen bridging), 9-arylthioxanthenes (sulfur bridging), 9, 10-dihydro-9-arylacridines (nitrogen bridging), 9-arylfluorenes (bridging through carbon–carbon covalent bond) have recently been the subject of number studies. Various applications of CRT molecules have been developed out of which application as photoredox catalyst is undoubtedly the most important. In this chapter, we have highlighted recent development of various CRT molecules, their photophysics, photochemistry and an application in the field of photoredox catalysis.

**Keywords:** conformationally restricted triarylmethanes (CRTs), 9-arylfluorenes, 9-arylxathenes, 9-arylthioxanthenes, photophysics, photochemistry, photoredox catalysis

## 1. Introduction

Triarylmethanes are compounds in which a carbon atom is linked to three aryl rings (both aromatic as well as heteroaromatic) which may be same or different. The most simple triarylmethane namely, triphenylmethane was first synthesized by August Kekule in 1872 by heating diphenyl mercury and benzal chloride [1]. Currently, more than thousand references can be found citing this molecule [2]. The development in the synthesis as well as application of these molecules have attracted various scientists with diverse field of research on this molecular scaffold leading to further high-end applications of these molecules. For example, simple triarylmethane derivatives have shown significant bio-activity against intestinal helminthes, filariae, trichomonads and trypanosomes [3]. Moreover, hydroxy substituted triarylmethanes are known for their antioxidant properties, antitumor activities as well as inhibitors of histidine protein kinases [4, 5]. Letrozole, Vorozole are effective Non-Steroidal Aromatase Inhibitors used commercially for the treatment of breast cancers [6–10]. (**Tram-3**), (**Tram-4**) are common organic dyes based on triarylmethane used in industry (vide **Chart 1**). Further, (**Tram-1**), (**Tram-2**), are common acid–base indicators [11], (**CRT-2**) and (**CRT-10**) are triarylmethane derivatives widely used as photoredox catalysts to synthesize various useful molecules (vide **Chart 2**) [12–14]. Various thermosetting condensed polynuclear aromatic resins, polymers, materials, drugs are also prepared based on this molecular scaffold (**Figure 1**) [15].

Through various non-covalent interactions, extended networks of hydrogen bonds, steric interactions, cycles/polycycles present in a molecule, the rotation of groups attached to the central carbon are restricted thereby decreasing the conformational mobility enjoyed by the molecules. This change in the conformation of the molecule in general not only modulates different physical properties of the molecule but also influences different photophysical and photochemical properties of the molecule [16, 17]. For example in open chain molecules due to rapid rotation of various bonds the vibrational relaxation of these molecules from the excited state to the ground state through non-radiative pathways are very rapid and these

**Phenol Dyes**

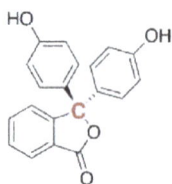

Phenolphthalein (**Tram-1**)    Bromocresol green (**Tram-2**)

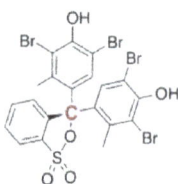

**Malachite green dyes**

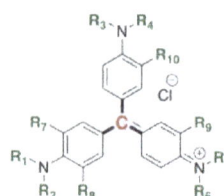

Malachite green (**Tram-3**)    Brilliant green (**Tram-4**)

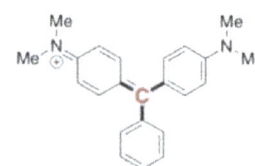

**Methyl Violet**

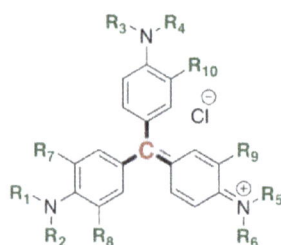

Methyl violet 2B ($R_1$ - $R_4$ = $CH_3$, $R_5$ - $R_{10}$ = H) (**Tram-5**).

Methyl violet 6B ($R_1$ - $R_5$ = $CH_3$, $R_5$ - $R_{10}$ = H) (**Tram-6**)

Methyl violet 10B ($R_1$ - $R_6$ = $CH_3$, $R_7$ - $R_{10}$ = H) (**Tram-7**)

Methylviolet ($R_1$ = $R_3$ = $R_3$ = $R_1$ - $R_4$ = $CH_3$, $R_5$ - $R_{10}$ = H) (**Tram-8**)

**Fuchsine dyes**

Pararosaniline ($R_3$ = $R_4$ = $CH_3$, $R_1$ = $R_2$ = $R_5$ - $R_{10}$ = H) (**Tram-9**)

Fuchsine acid ($R_1$ - $R_6$ = H, $R_7$ = $CH_3$, $R_8$ = $R_{10}$ = $SO_3Na$), $R_9$ = $SO_3^-$ (**Tram-10**)

New fuchsine (as chloride) ($R_1$ = $R_3$ = $R_3$ = $R_5$ = $R_7$ - $R_{10}$ = $CH_3$, $R_2$ = $R_4$ = $R_6$ = H) (**Tram-11**)

**Chart 1.** Some common conformationally *flexible* triarylmethanes molecules.

**9-Arylxanthene dyes**

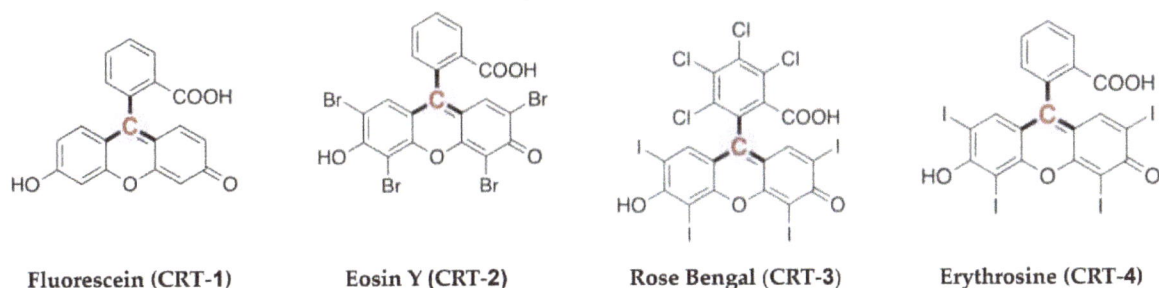

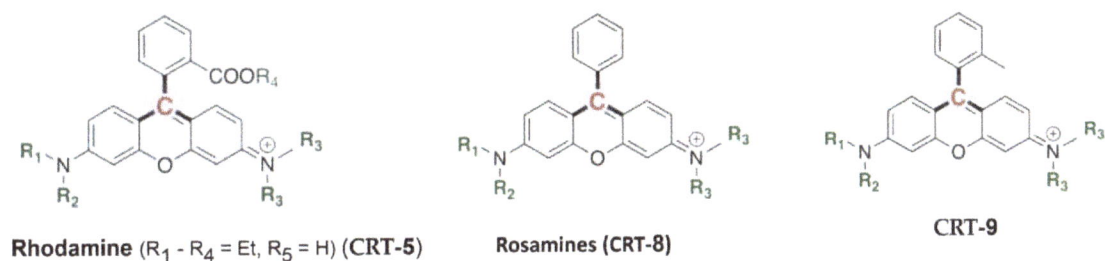

Fluorescein (CRT-1)          Eosin Y (CRT-2)          Rose Bengal (CRT-3)          Erythrosine (CRT-4)

Rhodamine (R₁ - R₄ = Et, R₅ = H) (CRT-5)          Rosamines (CRT-8)

Rhodamine 6G (R₁ = R₃ = Et, R₂ = R₄ = R₅ = H) (CRT-6)

Rhodamine 123 (R₁ = R₃ = H, R₅ = CH₃) (CRT-7)

**9-Aryl acridinium ions**                                    **Group 14 CRTs**

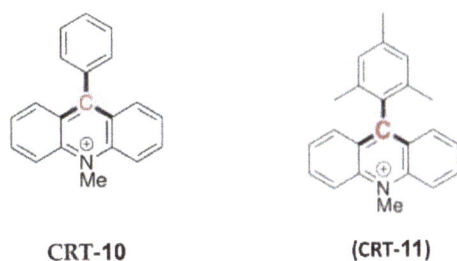

CRT-10          (CRT-11)          X=SiMe₂, (CRT-12), X=GeMe₂          X= S, (CRT-14)
                                   (CRT-13)                              X= Se, (CRT-15)

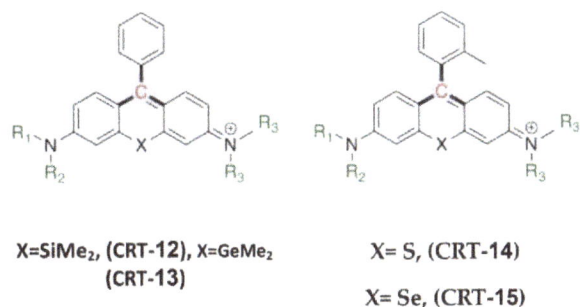

**Chart 2.** Some common conformationally *restricted* triarylmethanes molecules.

compounds are seldom fluorescent while in cyclic molecules due to the restriction of free rotation in the molecules there is a decrease in the vibrational relaxation through non-radiative pathways and hence there is a increase in the fluorescent quantum yield. To study the effect of this decrement in the conformational freedom and the properties that arise due to this restriction scientists term the molecules with such reduced conformational mobility as "conformationally restricted analogues" [18].

Due to the various non bonded interactions among the ortho protons/substituents in triaryl-methanes, the triarylmethanes exhibit 'molecular propeller conformation' in the ground state [17–21]. However, the three aryl rings attached to the central carbon atom rotate freely. This free rotation is restricted as shown in **Figure 2**, on bridging the two aryl rings with heteroatoms or bridging the aryl rings through bonds forming various types of molecules like 9-arylxanthenes (oxygen bridging) [22], 9-arylfluorenes (bridging through C-C bond) [23].

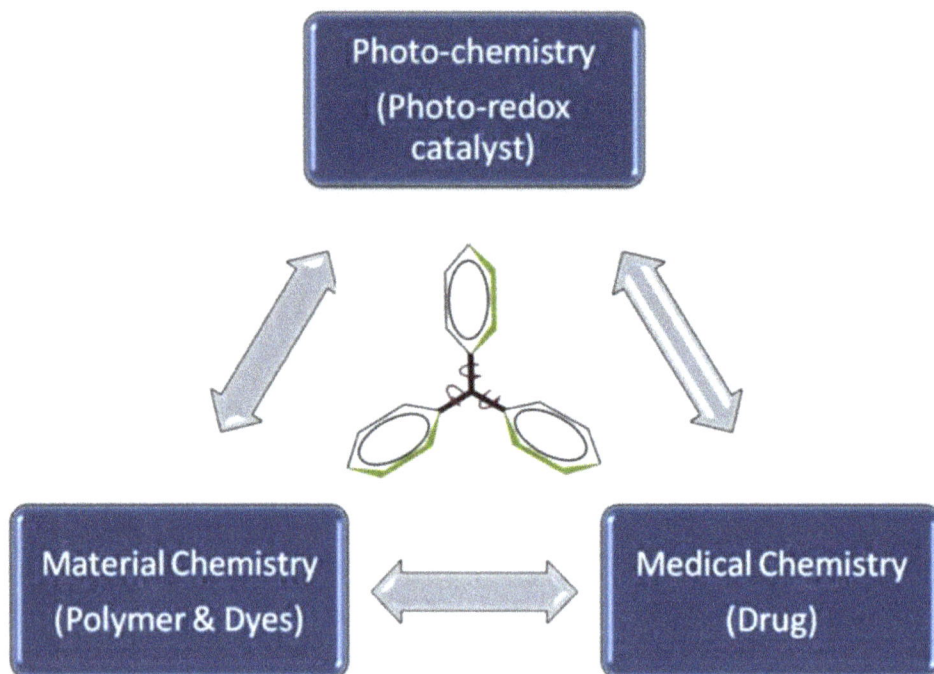

**Figure 1.** Application of triarylmethane molecular scaffold can be found in various molecular fields.

**Figure 2.** Restricting conformational flexibility through covalent bonding or bulky substitution to produce CRT molecules.

The presence of three aryl rings as well as the non-fluxional nature of the molecule results in easy abstraction of the corresponding methine hydrogen and thus faster generation of the corresponding free radicals; carbocations as well as carbanions [24]. In fact, Arnett and coworkers have described 9-arylxanthenes (a typical example of CRTs) as a subset of triarylmethanes [25, 26]. Both triarylmethanes, and 9-arylxanthenes are therefore can be considered as amphihydric compounds. These CRT compounds are reported to exist in two forms; one benzenoid and other quinoid structures due to this amphihydric nature as shown in **Figure 3** [27].

This benzenoid and quinoid structures along with the conformational restriction results in various interesting photophysical properties namely, the benzenoid form in the 9-arylxanthene derivatives being colorless while the quinoid form is intensely colored. In the benzenoid form the π electron delocalization in the chromophore is interrupted causing the absorption to be in the ultraviolet region and hence colorless while in the quinoid system this is uninterrupted causing the absorption to be in the visible region and hence is colored [28]. In the presence of base in (**Tram-1**), (**Tram-2**) and their derivatives intramolecular lactonization occur

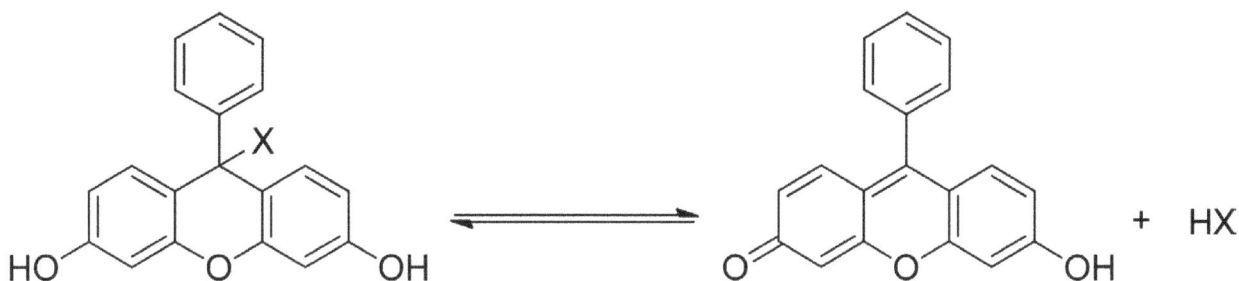

**Figure 3.** Two structural forms of CRTs.

thus reverting to a benzenoid structure and hence becoming colorless, which has been aptly used to develop various pH indicators as well different dye stuffs [29]. Moreover, in the CRTs, due to the rigid structure, the rate and extent of different relaxation processes involved in the relaxation of the excited molecules to the ground state differs thus showing different fluorescence quantum yield. For example, the CRT molecule (**CRT-1**), displays very high fluorescence (fluorescence quantum yield = 0.92, $\lambda$ excitation = 485 nm) whereas the corresponding flexible analogue, (**Tram-1**) is almost non-fluorescent [30].

## 2. Photophysical studies of CRTs: effect of conformational restriction on the photophysical properties

### 2.1. Photophysics of unrestricted triarylmethanes

The UV–Vis spectrum of the conformationally flexible triarylmethanes are very interesting with predominately two types of absorption bands being observed namely, the x band and the y band as shown (**Figure 4**). The x band arises through the transition of electron from the nonbonding molecular orbital of the molecule to the lowest unoccupied molecular orbital while the y band arises through the transition of the electron from the second highest occupied molecular orbital of the dye to the lowest unoccupied molecular orbital of the dye. This UV–Vis spectrum is found to depend strongly depends on various factors like the structure of the molecule [31–35], concentration [36], pH of the solution [37] as well as temperature [38]. Further due to the conformational flexibility and synchronous rotation among the attached aryl rings, the relaxation of the vibronically excited triarylmethanes to the ground state can occur through various radiation less processes. Due to this vibrational cascading the luminescence intensity and lifetime decrease in conformationally flexible triarylmethanes. However, this vibrational cascading is strongly affected by various factors like concentration, presence of other molecules (proteins/polymers, etc.), pH of the solution, as well as temperature thereby affecting the intensity of various luminescent processes [39].

### 2.2. Photophysics of conformationally restricted triarylmethanes

Decreasing the synchronous rotation of the aryl rings and hence the conformational flexibility in triarylmethanes has a profound influence in the observed photophysics of the triarylmethanes. The most significant change observed is the increase in the intensity of various luminescence processes. This is due to the fact that on conformational restriction, the relaxation of the excited

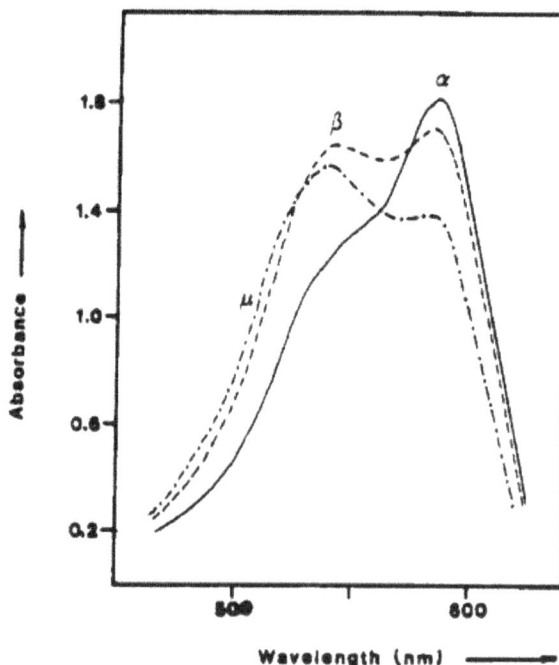

**Figure 4.** The absorption spectra of triarylmethane (Tram-5) showing the $\alpha$, $\beta$ and also the $\mu$ band at three different concentrations. Taken from Chem. Rev., 1993, 93, 381 with permissions from American Chemical Society.

state molecule to the ground state through various radiation less transitions decreases causing an increase in the intensity of the various luminescent processes [30]. Furthermore, the rate of the intersystem crossing also increases [40] thereby increasing the population of the triplet state [41].

### 2.2.1. Effect of structure

The functional groups/ the substituent atoms attached to a molecule modulate the photophysical properties of the molecule to a large extent. For example electron withdrawing groups like nitro groups affects the intramolecular charge transfer processes in the molecule [42], a tertiary amine group increase the rate of relaxation to the ground state through various non-radiative transitions arising due to the rapid rotation around the substituents attached to nitrogen and hence causes a decrease in fluorescence quantum yield [43, 44], heavy atom substituents like iodo, bromo increase the rate of intersystem crossing [45] through efficient spin-orbit coupling decreasing the fluorescence lifetime and quantum yield and concomitantly increasing the triplet emission yield [44].

In the CRTs the substituted aryl rings have a strong influence in the absorption-emission spectra of these compounds. For example depending on the pH of the solution the hydroxy xanthene dyes like (**CRT-1**), (**CRT-2**), (**CRT-3**) and (**CRT-4**), remain either in the protonated form or in the anionic form and the dianionic form. (**Figure 5**) [46], At low pH the protonated form predominates, while at higher pH the anionic form predominates. At still higher pH the dianionic form predominates. In the hydroxy xanthene dyes the emission quantum yield of the protonated form is the lowest while the emission quantum yield of the dianionic form is the highest [47]. Moreover, the greater conjugation observed in the dianionic form causes a red shift in the absorption as well as emission spectra. The intermediate mono anionic species forms a contact ion pair interaction and hence has a blue shift in the absorption-emission spectra [47].

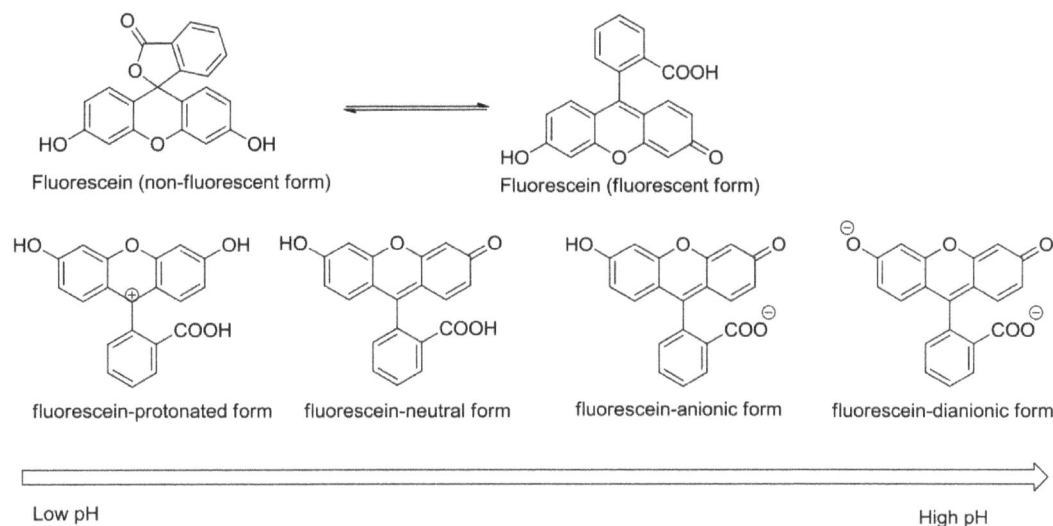

**Figure 5.** Fluorescent and non-fluorescent form of **CRT-1** at different pH.

### 2.2.2. Effect of solvent, concentration and temperature

The different structural forms of the hydroxy xanthene dyes have different hydrogen bonding ability with the solvent molecules which is reflected by the change in the UV–Vis absorption spectra on changing the hydrogen bonding ability of the solvent. For example, as observed by Martin and coworkers, with the increase in the hydrogen bonding effect of the solvent both the absorption as well as the fluorescence spectra of hydroxy xanthene dyes shows a blue shift [48]. This is due to the fact that hydrogen bonding interactions between the dye molecule and the solvent stabilizes the ground state of the hydroxy xanthene molecules more than the excited state causing an increase in the HOMO–LUMO energy gap and hence a blue shift in the UV–Vis absorption spectra is observed (**Figure 6**). Also the fluorescence quantum yield of these hydroxy xanthene dyes increase in such hydrogen bonding solvents. Moreover, in polar aprotic solvents in which no solute solvent interaction can take place through hydrogen bonding a red shift in the absorption-emission spectra is observed [49].

The hydrogen bonding interactions between the solute and the solvent molecules also stabilizes the singlet as well as the triplet energy states of the solute molecules with the solvent stabilization energy being most in the singlet ground state of the solute followed by the first excited singlet state. The triplet excited state, is least stabilized through hydrogen bonding. Thus the energy gap between the singlet excited state and the triplet excited state decrease in the hydrogen bonding solvent thereby increasing the rate of intersystem crossing. This interaction also organizes the dye molecules in a particular order changing the distortion required for internal conversion. This influences the rate constants as well as quantum yields for the internal conversion [50, 51].

Various non-covalent interactions between the dye molecule and the solvent as well as between the dye molecules result in the formation of dimer and/or higher order aggregates. These aggregates modulate the absorption-emission spectra CRTs strongly depend on the concentration of the CRT molecules. In fact, at higher concentration, the fluorescence spectra of many CRT molecules like **CRT-1, CRT-2** is self quenched [52]. For example dilute solution of (**CRT-1**) and (**CRT-2**) (concentration less than $5 \times 10^{-5}$ M) is in the monomer state

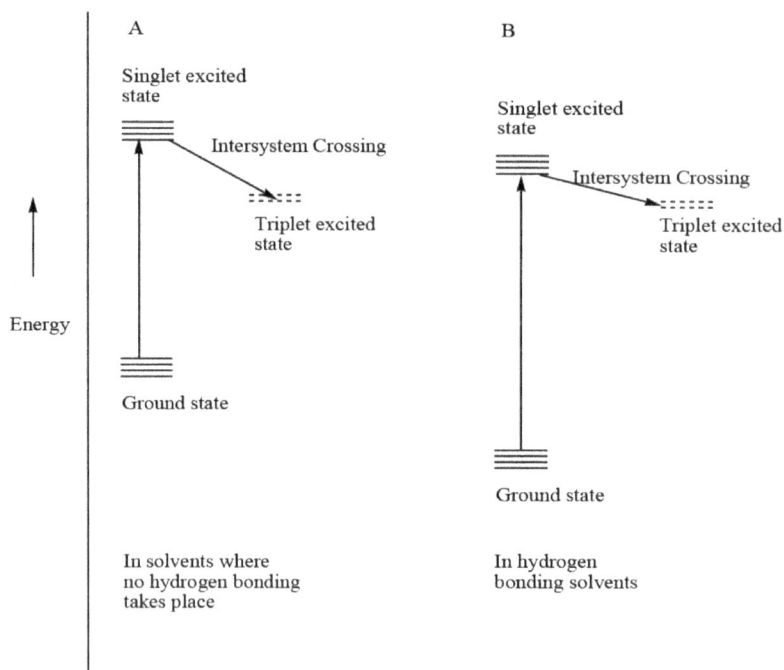

**Figure 6.** Pictorial representation of the energy levels of (**CRT-1**): (A) in solvents where no hydrogen bonding takes place. (B) In hydrogen bonding solvents. Figure not drawn to scale.

and displays the fluorescence spectra of the monomer form while at concentration range $5 \times 10^{-5}$ M to $10^{-3}$ M the fluorescence maxima shifts toward higher wavelength (smaller energy). At very high concentration (above $2 \times 10^{-3}$ M) with the concentration change there is no change in the fluorescence maxima [52].

Temperature has a profound influence on the fluorescent as well as phosphorescent properties of the CRT molecules [53–56]. This is not only due to the dependence of the population of molecules in a particular electronic state in a molecule depends on the temperature of the molecule (according to Boltzmann distribution factor) but also due to the change in the probability of de-excitation from various vibrational levels through non-radiative transitions changing the fluorescent quantum yield on change in the temperature. Moreover, at low temperature, **CRT-5**, **CRT-6**, **CRT-7** show a decrease in the rate of photo degradation showing a strong dependence of the photophysical properties on temperature as shown by Nagano and coworkers [57].

### 2.2.3. Photoinduced electron transfer (PET) in CRTs

Excitation of molecules from the ground state to the excited state increases the redox activity of the compound. This is due to the fact that on excitation of an electron from the ground state creates an "electron hole" in its highest occupied molecular orbital as well as equivalent amount of electron occupancy in the lowest unoccupied molecular orbital. However, the electrons in the LUMO energy level are loosely bound and hence can be easily detached (hence greater reducing ability) while the electron hole in the HOMO results in greater electron affinity (hence greater oxidizing ability).

The fluorescence quantum yield of the 9-arylxanthenium ion type of CRT derivatives have been found to be high (fluorescence $\Phi$= 0.45) [53]. However, this high fluorescence quantum yield is effectively quenched in various CRTs molecules having electron-rich aromatic substituents due to a PET from the electron-rich aryl rings to the fluorophore takes place, quenching the observed fluorescence [54]. For example the fluorescence quantum yield of (**CRT-6**) is 0.90 while when the carboxy group of (**CRT-6**) is replaced with an electron donating methyl group (**CRT-9**) the fluorescence quantum yield of the resulting compound reduces to 0.35 [55]. Similarly, the fluorescence quantum yield of (**CRT-1**) is 0.92 while the derivative of (**CRT-1**) formed by adding an electron-rich amino group in the 4′ position has fluorescence quantum yield of 0.015 [56]. This effect has been proved through direct flash photolysis studies as well as through directly isolating the radical/radical ion species. Moreover, this fluorescence quenching strongly depends on the electronic nature of the aromatic donors, the oxidizing potential of the aromatic donors, as well as on the orientation of the electron donors. Thus O-arylated derivatives of (**CRT-1**) have very low fluorescent quantum yield compared to (**CRT-1**) molecules and on oxidizing with highly reactive oxygen species revert back to the (**CRT-1**) molecule thus increasing the fluorescence quantum yield (**Scheme 1**) [57]. However in molecules with electron poor aromatic substituents this photoinduced electron transfer process is shut down due to the electron deficient aryl rings and hence CRTs with electron poor aryl rings has higher fluorescence quantum yield compared to the CRTs with electron-rich substituents. Moreover, due to higher excitation energy of the singlet state the singlet excited state is more oxidizing compared to the triplet excited state [58].

### 2.2.4. Photophysics of molecules formed through replacement of the bridging atom of CRT with Si atom and other groups 14 and group 16 elements

Molecules formed through the replacement of the bridging heteroatoms in the CRT molecules with Si atom or other groups 14 elements such as Ge, Sn, modulate the energy levels of the parent CRT molecules through various stereoelectronic effects causing a high bathochromic shift in the U.V-Vis absorption spectra. This is because the substituents attached to the group-14 atom has a strong electron pushing inductive effect as well as has a strong $\sigma^*$-$\pi^*$ conjugation between the group 14 atom substituent $\sigma^*$ orbital and $\pi^*$ orbital of the fluorophore modulating the HOMO as well as LUMO energy levels. Moreover, this bathochromic shift in the absorption maxima among the group 14 substituted CRTs decreases down the

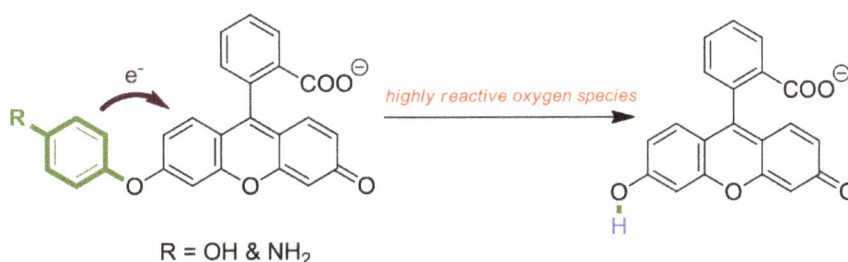

R = OH & NH$_2$

**Scheme 1.** Photoinduced electron transfer between aryl rings and electron deficient xanthene rings.

group opposite to that observed when the oxygen atom is replaced by group 16 elements. This is due to the fact that in case of group 16 elements the resonance effect takes place between the lone pair of electrons in the chalcogens and the fluorophore with the positive charge being efficiently delocalized throughout the fluorophore resulting in a concomitant decrease in the energy gap between the frontier orbitals. For example, the absorption maxima, fluorescence maxima, fluorescence quantum yield for (CRT-8) are respectively 552 nm, 575 nm and 0.84 whereas the same for (CRT-14) and (CRT-15) formed through the replacement of the bridging oxygen atom with S and Se are 571 nm ($\lambda_{abs}$ CRT-14), 582 nm ($\lambda_{abs}$ CRT-15), 599 nm ($\lambda_f$ CRT-14), 608 nm ($\lambda_f$ CRT-15), 0.44 ($\Phi_f$ CRT-14) and 0.009 ($\Phi_f$ CRT-15) respectively. The absorption maxima, fluorescence maxima, fluorescence quantum yield for (CRT-9) are 549 nm, 575 nm and 0.35 respectively. For compound (CRT-12) and (CRT-13) formed through the replacement of the bridging oxygen atom with Si and Ge the absorption maxima, fluorescence maxima and fluorescence quantum yield are 646 nm ($\lambda_{abs}$ CRT-12), 635 nm ($\lambda_{abs}$ CRT-13), 660 nm ($\lambda_f$ CRT-12), 649 nm ($\lambda_f$ CRT-13), 0.31 ($\Phi_f$ CRT-12) and 0.34 ($\Phi_f$ CRT-13)respectively [55].

# 3. Photochemistry of CRTs

## 3.1. Photooxidation and photoreduction

CTRs with photocleavable appendages dissociate into corresponding cations on irradiation of ultraviolet light. The corresponding ions display a quinoid structure (vide supra, **Figure 2**) and are highly colored. The resulting carbocation takes part in various photooxidation or photoreduction processes. The photooxidation process involves ejection of an electron from the excited state to form the respective triarylmethane carbocation (presumably in the triplet state) which may further react with molecular oxygen to form organic peroxides. It may proceed to further oxidized products or may lose an electron to the solvent molecule to form the respective triaryl-methane radical ion or solvated electrons which further reacts with molecular oxygen to form the peroxides (vide **Scheme 2**). The photoreduction process involves either hydrogen atom abstraction or electron abstraction process through a photoexcited triplet state.

**Radical dication pathway**

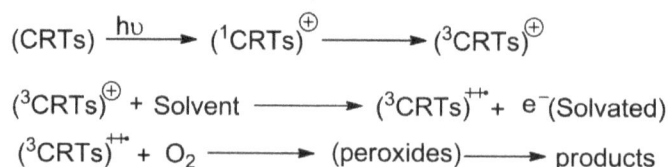

$$(CRTs) \xrightarrow{h\upsilon} (^1CRTs)^{\oplus} \longrightarrow (^3CRTs)^{\oplus}$$

$$(^3CRTs)^{\oplus} + Solvent \longrightarrow (^3CRTs)^{++} + e^-(Solvated)$$

$$(^3CRTs)^{++} + O_2 \longrightarrow (peroxides) \longrightarrow products$$

**Radical pathway**

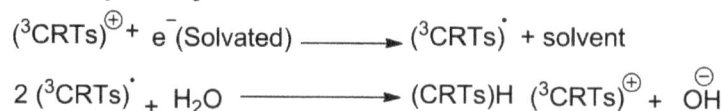

$$(^3CRTs)^{\oplus} + e^-(Solvated) \longrightarrow (^3CRTs)^{\cdot} + solvent$$

$$2 (^3CRTs)^{\cdot} + H_2O \longrightarrow (CRTs)H (^3CRTs)^{\oplus} + \overset{\ominus}{OH}$$

Scheme 2. Different photooxidation processes of CRTs.

## 3.2. Photoredox catalysis

For practitioners of synthetic chemistry in the pharmaceutical industry, materials industry as well as in the academia the ultimate goal is to develop practical, scalable processes with minimum impact on the environment. On this line, if different sources of energy available to humanity are compared then the light is the most abundant, endless, renewable, clean form of energy and using light to transform raw materials to value-added products will always be of high demand. Moreover, this mode of catalysis provides various new reactions through the facile generation of reactive intermediates which are otherwise difficult/ impossible to obtain. Though transition metal chromophores were used initially as photoredox catalysts, recently organic dyes are being used.

## 3.3. CRTs as photoredox catalysts

The rich photophysics of CRT derivatives (e.g., facile intersystem crossing due to heavy atom effect, photooxidation, photoreduction, PET, etc.) allow these molecules to be efficiently used as photocatalysts. Especially, the hydroxy xanthene dyes (**CRT-2**), (**CRT-3**), (**CRT-10**) and (**CRT-11**) are noteworthy.

### 3.3.1. (CRT-2) as a Photoredox catalyst

Due to the heavy atom effect in the (**CRT-2**) molecule, rapid intersystem crossing to the triplet state can take place causing a high triplet quantum yield [59]. Moreover, a photo-induced electron transfer from the benzoate ring to the xanthene fluorophore results in efficient modulation of the redox properties if the dye molecule. The redox potentials of the triplet excited state of (**CRT-2**) indicates that it becomes more oxidizing as well as reducing on photoexcitation, thus photoexcited (**CRT-2**) can act as an electron donor (reducing property) or electron acceptor (oxidizing property) under suitable conditions. Another property of this molecule that is worth mentioning is the fact that due to significant triplet quantum yield (**CRT-2**) can act as triplet energy transfer agent and thus can generate singlet oxygen from air. [60].

### 3.3.2. Use of (CRT-2) in reduction reactions

### 3.3.2.1. Nitrobenzene to aniline

Under green light irradiation and in the presence of sacrificial reducing agents like triethanolamine, strong electron acceptors like nitrobenzenes are reduced to aniline derivatives (**Scheme 3**) [61]. Through flash photolysis experiments the mechanism of the reaction was established. (**CRT-2**) loses an electron to the strong electron acceptor nitro group which is converted to the intermediate radical anion. The photocatalyst returns to the ground state through an electron acceptance from the sacrificial reductant triethanolamine which is converted to the aminyl radical cation. This radical cation intermediate then reduces the nitro intermediate radical anion to aniline derivative.

**Scheme 3.** (**CRT-2**) catalyzed photoredox conversion of nitrobenzene to aniline.

### 3.3.2.2. Desulfonylation

A metal-free green protocol for the removal of sulfonyl group was reported by Wu and coworkers (**Scheme 4**) [62]. Blue light irradiation to bis tetrabutylammonium (**CRT-2**) salt excites it to its excited state which is oxidatively quenched by β-arylketosulfones resulting the (**CRT-2**) radical cation and the radical anion of β-arylketosulfones which then undergoes desulfonylation to the aryl ketone radical. The sacrificial reductant closes the catalytic cycle by reducing the (**CRT-2**) radical cation to (**CRT-2**) photocatalyst and its radical cation which releases a hydrogen atom to the ketone radical.

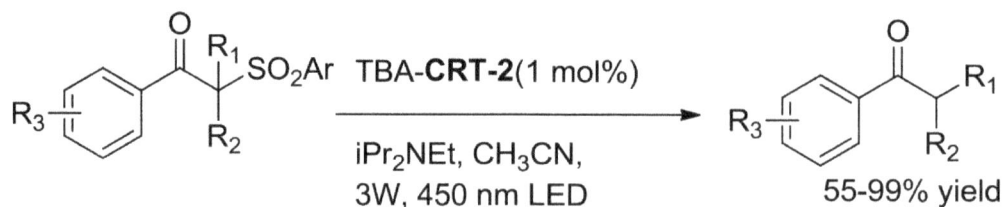

**Scheme 4.** Reductive desulfonylation of β-keto sulfones catalyzed by **CRT2**.

### 3.3.3. Use of (CRT-2) in oxidation reactions

### 3.3.3.1. Benzylic oxidations

In a formal Kornblum oxidation, when benzyl bromides are heated at 80°C in the presence of (**CRT-2**) as the photocatalyst in the presence of oxygen in DMSO solvent moderate to high amount of arylaldehydes are formed (**Scheme 5**) [63]. However, interestingly, benzyl chlorides were inert to this oxidation reaction. Moreover, both oxygen, as well as DMSO, was found to be essential for the success of the reaction.

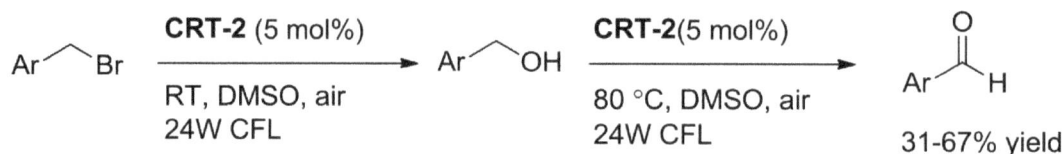

**Scheme 5.** (**CRT-2**) catalyzed conversion of benzyl bromides to benzaldehydes.

### 3.3.3.2. Oxidative iminium ion formation

Single electron transfer from tertiary amines to the excited state of the (**CRT-2**) leads to the formation of aminyl radical cation and the radical anion of eosin. The radical anion of (**CRT-2**) further loses an electron to oxygen molecule to form the reactive superoxide radical anion which further abstracts a proton from the aminyl radical cation to form the imine intermediate resulting in the oxidative imminium ion formation [64]. The high electrophilicity of the iminium ion can be further exploited for the efficient construction of C-C and C-P bonds by treating the formed iminium ions with the nucleophilic partners like indole, dialkyl malonates, dialkyl phosphonates, etc. (**Scheme 6**).

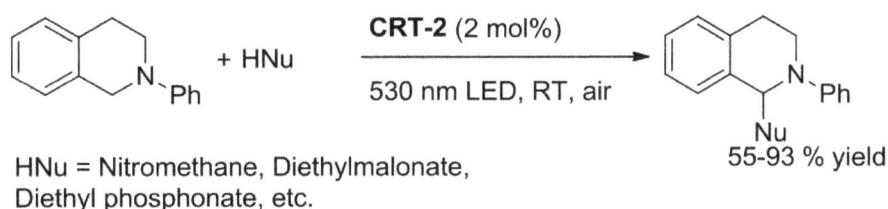

HNu = Nitromethane, Diethylmalonate,
Diethyl phosphonate, etc.

**Scheme 6.** C-C and C-P bond formation through oxidative iminium ion formation.

### 3.3.3.3. Benzylic bromination

Using carbon tetrabromide as an efficient source of bromine, morpholine as a reducing agent and (**CRT-2**) as a photocatalyst, Tan and coworkers developed an efficient process for the selective bromination of aliphatic and benzylic C-H bonds (**Scheme 7**) [65]. The authors suggested an intermediate formation of N-morpholino radical for the crucial C-H functionalization step.

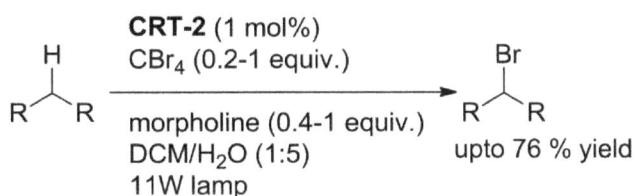

**Scheme 7.** Photoredox C-H bromination of alkanes catlyzed by (**CRT-2**).

### 3.3.4. Use of (CRT-2) in redox neutral transformation

### 3.3.4.1. Arylation reaction using aryldiazonium salts

Excited (**CRT-2**) molecules take part in a SET with aryldiazonium ions to form aryl radicals and (**CRT-2**) radical cation. The formed aryl radical then adds to the double bond of the heteroarenes to form the corresponding radical intermediate which further undergoes a SET to either the (**CRT-2**) radical cation to close the catalytic cycle or to the aryldiazonium ions to form aryl radicals. The corresponding carbocations so formed further aromatizes to give the C-H arylated product. Thus aryldiazonium salts can be used as a source of aryl radicals for redox neutral arylation (**Scheme 8**).

**Scheme 8.** Redox neutral arylation of heteroarenes.

### 3.3.4.2. (CRT-11) as photoredox catalyst

The, (**CRT-11**) a class of triarylmethane derivatives, with electron-rich aryl rings in the 9th position, display a unique ability to form a donor-acceptor type of dyad where the electron-rich aromatic ring acts as electron donor while the acridinium ring acts as an electron acceptor. Moreover, the presence of sterically hindered aryl group in the 9th position shields the benzylic position from any nucleophilic attack and also shifts the absorption maxima to the visible range through electron delocalization with the aryl ring. When irradiated with light of suitable frequency (approx. 450 nm absorption maxima) an intramolecular electron transfer from the electron-rich aromatic nucleus to the singlet excited state of the acridinium ring takes place (PET) leading to the formation of the long lived electron transfer excited state. As evident from the reduction potential of the excited state of this molecule, this molecule can easily accept electrons to oxidize organic substrates to its radical cations which can further react with nucleophiles to form the radical adducts while the dye molecule itself loses electrons to electron acceptors like oxygen to form hydroperoxide radical. This radical may further react with the adduct to form its hydroperoxide and further to its oxidized species and hydrogen peroxide (**Figure 7**). Thus in the presence of air (**CRT-11**) can act as an efficient oxidizing agent reducing oxygen molecule (in air) to hydrogen peroxide [13]. For example, this class of dyes has been

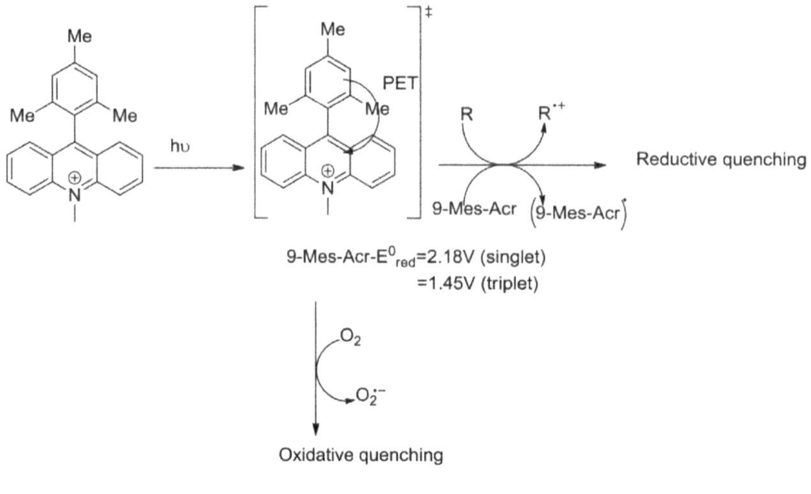

**Figure 7.** PET, reduction potential, reductive quenching and oxidative quenching of 9-Mes acridinium ions.

effective photocatalysts in the oxygenation of aromatic hydrocarbons to aromatic aldehydes, in C-H oxidation of cycloalkanes and also in the alkene hydrofunctionalization.

### 3.3.4.3. Use of 9-Mes acridinium ions as oxidizing agent

### 3.3.4.3.1. Oxidation of toluene to benzaldehyde

In the presence of (CRT-11) as photocatalyst and under oxygen atmosphere methyl arenes are converted to their respective aldehydes in high yields (Scheme 9) [66]. Remarkably in substrates with multiple methyl groups only single methyl group is oxidized thus showing high selectivity of the method. The acridinium photocatalyst oxidizes the arenes to the corresponding arene radical cation, which further yields a benzylic radical through loss of a proton. Aerial oxygen may oxidize the benzyl radical to peroxyl radical which disproportionate to the observed product. A SET from the oxygen molecule regenerates the acridinium ion catalyst to the ground state.

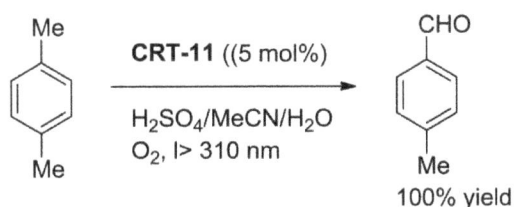

Scheme 9. Selective oxidation of p-xylene to p-tolualdehyde.

### 3.3.4.3.2. C-H oxidation of cycloalkanes

Fukuzumi and coworkers reported a photocatalytic process to oxidize the inert C-H bonds of cycloalkanes [67]. Using the photoredox catalyst (CRT-11) in the presence of oxygen and hydrochloric acid a mixture of cycloalkanols and ketones could be obtained (Scheme 10). Though low yields of the product were obtained, the high TON could be achieved. Moreover, the ability to functionalize an inert C-H bond is the utility of the process. The mechanism of the reaction involves oxidation of the chloride ions to chlorine radicals which abstract a hydrogen atom from the C-H bond of the cycloalkanes to form the alkyl radical. The alkyl radical reacts with oxygen to form alkyl peroxyl radical which disproportionate to the products. The involvement of the alkyl peroxyl radical is determined through EPR spectra.

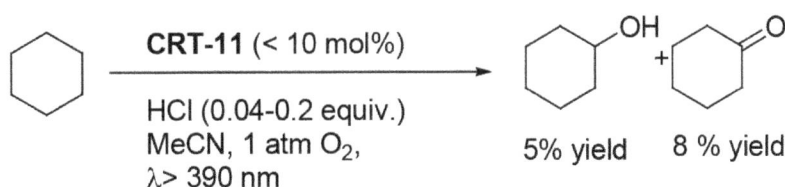

Scheme 10. Photoredox C-H oxidation of cycloalkanes.

*3.3.4.4. Use of (CRT-11) for alkene hydrofunctionalization reaction*

The regioselective alkene hydrofunctionalization is an essential yet challenging task in synthetic chemistry. Nicewicz developed an efficient protocol for alkene hydrofunctionalization using the (**CRT-11**) as the photocatalyst (**Scheme 11**) [68–70]. Thus in the presence of (**CRT-11**) alkenes can be oxidized to their corresponding radical cations, which react with the nucleophiles at the less substituted position to produce the more stable nucleophile cation radical adduct thus giving rise to the observed regioselectivity. Further, a HAT from added HAT to the produced radical gives the observed product and the corresponding radical of the HAT agent. This radical then regenerates the (**CRT-11**) ions catalyst in the ground state, the HAT agent being converted to its anion. The so produced anion of the HAT agent deprotonates the nucleophile cation radical adducts to regenerate the HAT agent. This transformation allows efficient construction of C-C, C-O, C-N, etc. bonds through the use of nucleophilic arenes, carboxylic acids, alcohols and amines. Of special note is the fact that through the use of photoredox catalyst, addition of hydrochloric acid can take place in anti Markovnikov fashion.

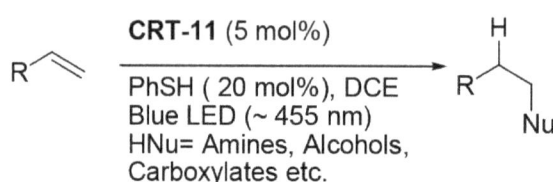

**Scheme 11.** (**CRT-11**) catalyzed alkene hydrofunctionalization.

# 4. Conclusion

The free rotation among the three aryl rings of the triarylmethane molecule is restricted on bridging the aryl rings with heteroatoms or through bonds. The resulting conformationally restricted triarylmethane molecules have very rich photophysics which is the topic of this chapter. This conformational restriction decreases the rate of relaxation of the excited molecules through vibrational cascading increasing the fluorescent intensity of the resulting compound. Heavy atom substituted CRTs shown a high rate of intersystem crossing thus increasing the triplet quantum yield and also gives rise to unique photochemistry which is discussed. This property is used currently to develop small organic molecules catalyzing the conversion of light energy to value-added products which is also discussed. We anticipate that this chapter will benefit readers interested to develop novel photocatalytic systems to synthesis various value-added products.

# Acknowledgements

SM gratefully acknowledges Science and Engineering Research Board, India for the award of National Postdoctoral Fellowship (File number: PDF/2016/001146). Authors also acknowledges department of chemistry, Institute of Science, Banaras Hindu University for providing infrastructural facilities.

## Conflict of interest

The authors declare no conflict of interest.

## Author details

Sankalan Mondal and Satyen Saha*

*Address all correspondence to: satyen.saha@gmail.com

Department of Chemistry, Institute of Science, Banaras Hindu University, Varanasi, India

## References

[1] Kekulé A, Franchimont A. Berichte der Deutschen Chemische Gesellschaft. 1872;**5**:906

[2] For a comprehensive overview on the synthesis and biological activity of triarylmethanes please see : Mondal S, Panda G. RSC Advances. 2014;**4**:28317

[3] Schnitzer RJ, Hawking F. Experimental Chemotherapy. Vol. I. New York, NY: Academic; 1963

[4] Yamato M, Hashigaki K, Yasumoto Y, Sakai J, Luduena RF, Banerjee A, Tsukagoshi S, Tashiro T, Tsuruo T. Synthesis and antitumor activity of tropolone derivatives. 6. Structure-activity relationships of antitumor-active tropolone and 8-hydroxyquinoline derivatives. Journal of Medicinal Chemistry. 1987;**30**:1897

[5] Mondal S, Roy D, Jaiswal M, Panda G. A green synthesis of unsymmetrical triarylmethanes via indium (III) triflate catalyzed friedel crafts alkylation of o-hydroxy bisbenzylic alcohols under solvent free conditions. Tetrahedron Letters. 2018;**59**:89

[6] Bossche HV, Willemsens G, Roels I, Bellens D, Moereels H, Coene MC, Jeune LL, Lauwers W, Janssen PA. R 76713 and enantiomers: Selective, nonsteroidal inhibitors of the cytochrome P450-dependent oestrogen synthesis. Biochemical Pharmacology. 1990;**40**:1707

[7] Recanatini M, Cavalli A, Valenti P. Nonsteroidal aromatase inhibitors: Recent advances. Medicinal Research Reviews. 2002;**22**:282

[8] Wood PM, Woo LL, Labrosse JR, Trusselle MN, Abbate S, Longhi G, Castiglioni E, Lebon F, Purohit A, Reed MJ. Chiral aromatase and dual aromatase–steroid sulfatase inhibitors from the letrozole template: Synthesis, absolute configuration, and in vitro activity. Journal of Medicinal Chemistry. 2008;**51**:4226

[9] Takahashi K, Yamagishi G, Hiramatsu T, Hosoya A, Onoe K, Doi H, Nagata H, Wada Y, Onoe H, Watanabe Y, Hosoya T. Practical synthesis of precursor of [N-methyl-11C]vorozole, an efficient PET tracer targeting aromatase in the brain. Bioorganic & Medicinal Chemistry. 2011;**19**:1464

[10] Wood PM, Woo LW, Thomas MP, Mahon MF, Purohit A, Potter BV. Aromatase and dual aromatase-steroid sulfatase inhibitors from the letrozole and vorozole templates. ChemMedChem. 2011;**6**:1423

[11] Rys P, Zollinger H. Dyes as indicators. Fundamentals of the Chemistry and Application of Dyes. New York, NY: Wiley-Interscience; 1972

[12] Hari DP, König B. Synthetic applications of eosin Y in photoredox catalysis. Chemical Communications. 2014;**50**:6688

[13] Fukuzumi S, Ohkuboa K. Selective photocatalytic reactions with organic photocatalysts. Chemical Science. 2013;**4**:561

[14] Romero NA, Nicewicz DA. Organic photoredox catalysis. Chemical Reviews. 2016;**116**:10075

[15] Ota M, Otani S, Kobayashi K. The preparation and properties of the condensed polynuclear aromatic (COPNA) resins using an aromatic aldehyde as crosslinking agent. Chemistry Letters. 1989;**18**:1175

[16] Dwivedi N, Kumar Panja S, Das M, Saha S, Sunkari SS. Anion directed structural diversity in zinc complexes with conformationally flexible quinazoline ligand: Structural, spectral and theoretical studies. Dalton Transactions. 2016;**45**:12053

[17] Panja SK, Dwivedi N, Saha S. Manipulating the proton transfer process in molecular complexes: synthesis and spectroscopic studies. Physical Chemistry Chemical Physics. 2016;**18**:21600

[18] Grygorenko OO, Radchenko DS, Volochnyuk DM, Tolmachev AA, Komarov I. Bicyclic conformationally restricted diamines. Chemical Reviews. 2011;**111**:5506

[19] Sabacky MJ, Johnson SM, Martin JC, Paul C. Steric effects in ortho-substituted triarylmethanes. Journal of the American Chemical Society. 1969;**91**:7542

[20] Gust D, Mislow K. Analysis of isomerization in compounds displaying restricted rotation of aryl groups. Journal of the American Chemical Society. 1973;**95**:1535

[21] Finocchiaro P, Gust D, Mislow K. Separation of conformational stereoisomers in a triarylmethane. Journal of the American Chemical Society. 1973;**95**:8172

[22] McKinley SV, Grieco PA, Young AE, Freedman HH. Rotational barriers and conformational studies in 9-arylxanthyl derivatives. Journal of the American Chemical Society. 1970;**92**:5900

[23] Chandross EA, Sheley CF Jr. Some 9-aryl fluorenes. Ring-current effects on nuclear magnetic resonance spectra, carbonium ions, and the 9-mesitylfluorenyl radical. Journal of the American Chemical Society. 1968;**90**:4345

[24] Vougioukalakis GC, Roubelakis MM, Orfanopoulos M. Radical reactivity of Aza[60] fullerene: preparation of monoadducts and limitations. The Journal of Organic Chemistry. 2010;**75**:4124

[25] Arnett EM, Flowers RA, Meekhof AE, Miller L. Energetics of formation for conjugate xanthyl carbenium ions, carbanions, and radicals by hydride, proton, and electron transfer in solution and their reactions to give symmetrical bixanthyls. Journal of the American Chemical Society. 1993;**115**:12603

[26] Arnett EM, Flowers R, Ludwig RT, Meekhof AE, Walek S. Triarylmethanes and 9-arylxanthenes as prototypes amphihydric compounds for relating the stabilities of cations, anions and radicals by C-H bond cleavage and electron transfer. Journal of Physical Organic Chemistry. 1997;**10**:499

[27] Gomberg M, West CJ. On triphenylmethyl. XXI. Quinocarbonium salts of the hydroxyxanthenols. Journal of the American Chemical Society. 1912;**34**:1529

[28] Jacob JA, Frantzeskos J, Kemnitzer NU, Zilles A, Drexhage KH. New fluorescent markers for the red region. Spectrochimica Acta, Part A: Molecular and Biomolecular Spectroscopy. 2001;**57**:2271

[29] Gessner T, Mayer U. Triarylmethane and Diarylmethane dyes. In: Ullmann's Encyclopedia of Industrial Chemistry. Weinheim: Wiley-VCH; 2005

[30] Boguta A, Wrobel D. J. Fluorescein and phenolphthalein—correlation of fluorescence and photoelectric properties. Fluores. 2001;**11**:129

[31] Gray GW, editor. Steric Effects in Conjugated Systems. London: Butterworth; 1958

[32] Barker CC. Steric Effects in Conjugated Systems. Chapter 4. p. 34

[33] Dewar MJS. Steric Effects in Conjugated Systems Chapter 5. p. 46

[34] Korppi-Tommola J, Kolehmainen E, Salo E, Yip RW. The temperature-dependent redshift of the visible absorption spectra of crystal violet in alcohol solutions. Chemical Physics Letters. 1984;**104**:373

[35] Barker CC, Bride MH, Stamp A. 796. Steric effects in di- and tri-arylmethanes. Part I. Electronic absorption spectra of o-methyl derivatives of Michler's hydrol blue and crystal violet; conformational isomers of crystal violet. Journal of the Chemical Society. 1959;**4**:3957

[36] Crenzi V, Quadriioglk F, Vikhano VJ. Interaction of crystal violet and polyanions in aqueous solution. Journal of Macromolecular Science Part A Pure and Applied Chemistry. 1967;**A1**:917

[37] Roes WJ, White JC. Application of pyrocatechol violet as a colorimetric reagent for tin. Analytical Chemistry. 1961;**33**:421

[38] Duxbury DF. The photochemistry and photophysics of triphenylmethane dyes in solid and liquid media. Chemical Reviews. 1993;**93**:381

[39] Murrel JN. The Theory of the Electronic Spectra of Organic Molecules. London: John Wiley and Sons Inc; 1963. p. 210

[40] Fleming GR, Knight AWE, Morris JM, Morrison RJS, Robinson GW. Picosecond fluorescence studies of xanthene dyes. Journal of the American Chemical Society. 1977;**99**:4306

[41] Shen T, Zhao ZG, Yu Q, Xu HJ. Photosensitized reduction of benzil by heteroatom-containing anthracene dyes. Journal of Photochemistry and Photobiology A: Chemistry. 1989;**47**:203

[42] Panja SK, Dwivedi N, Saha S. Tuning the intramolecular charge transfer (ICT) process in push–pull systems: effect of nitro groups. RSC Advances. 2016;**6**:105786

[43] Saha S, Samanta A. Photophysical and dynamic NMR studies on 4-Amino-7-nitrobenz-2-oxa-1, 3-diazole derivatives: elucidation of the nonradiative deactivation pathway. The Journal of Physical Chemistry. A. 1998;**102**:7903

[44] Saha S, Samanta A. Influence of the structure of the amino group and polarity of the medium on the photophysical behavior of 4-Amino-1,8-naphthalimide derivatives. The Journal of Physical Chemistry. A. 2002;**106**:4763

[45] Lakowicz JR. Principles of Fluorescence Spectroscopy. 3rd ed. New York: Springer; 2006

[46] Markuszewski R, Diehl H. The infrared spectra and structures of the three solid forms of fluorescein and related compounds. Talanta. 1980;**27**:937

[47] McQueen PD, Sagoo S, Yao H, Jockusch RA. On the intrinsic photophysics of fluorescein. Angewandte Chemie, International Edition. 2010;**49**:9193

[48] Martin MM. Hydrogen bond effects on radiationless electronic transitions in xanthene dyes. Chemical Physics Letters. 1975;**35**(1):105

[49] Volman DH, Hammond GS, Neckers DC, editors. Advances in Photochemistry. Vol. 18. Hoboken, New Jersey, USA: John wiley and Sons

[50] Xu D, Neckers DC. Aggregation of rose bengal molecules in solution. Journal of Photochemistry. 1987;**40**:361

[51] Aguilera OV, Neckers DC. Aggregation phenomena in xanthene dyes. Accounts of Chemical Research. 1989;**22**:171

[52] Arbeola L. Flourescence self-quenching of Halofluorescein dyes. Journal of Photochemistry. 1982;**18**:161

[53] Melhuish WH. Quantum efficiencies of fluorescence of organic substances: effect of solvent and concentration of the fluorescent solute. The Journal of Physical Chemistry. 1961;**65**:229

[54] Samanta A, Gopidas KR, Das PK. Electron acceptor behavior of 9-phenylxanthenium carbocation singlet. Chemical Physics Letters. 1990;**167**:165

[55] Koide Y, Urano Y, Hanaoka K, Terai T, Nagano T. Evolution of group 14 rhodamines as platforms for near-infrared fluorescence probes utilizing photoinduced electron transfer. ACS Chemical Biology. 2011;**6**:600

[56] Munkholm C, Parkinson DR, Walt DR. Intramolecular fluorescence self-quenching of fluoresceinamine. Journal of the American Chemical Society. 1990;**112**:2608

[57] Setsukinai K, Urano Y, Kakinuma K, Majima HJ, Nagano T. Development of novel fluorescence probes that can reliably detect reactive oxygen species and distinguish specific species. The Journal of Biological Chemistry. 2003;**278**:3170

[58] Johnston LJ, Wong DF. Electron transfer reactions of triplet 9-arylxanthenium and 9-arylthioxanthenium cations. The Journal of Physical Chemistry. 1993;**97**:1589

[59] Penzkofer A, Beidoun A, Daiber M. Intersystem-crossing and excited-state absorption in eosin Y solutions determined by picosecond double pulse transient absorption measurements. Journal of Luminescence. 1992;**51**:297

[60] Redmond RW, Gamlin JN. A Compilation of singlet oxygen yields from biologically relevant molecules. Photochemistry and Photobiology. 1999;**70**:391-475

[61] Yang X-J, Chen B, Zheng L-Q, Wu L-Z, Tung C-H. Highly efficient and selective photocatalytic hydrogenation of functionalized nitrobenzenes. Green Chemistry. 2014;**16**:1082

[62] Yang D-T, Meng Q-Y, Zhong J-J, Xiang M, Liu Q, Wu L-Z. Metal-free desulfonylation reaction through visible-light photoredox catalysis. European Journal of Organic Chemistry. 2013;**2013**:7528

[63] Kornblum N, Jones WJ, Anderson GJ. A new and selective method of oxidation. The conversion of alkyl halides and alkyl tosylates to aldehydes. Journal of the American Chemical Society. 1959;**81**:4113

[64] Hari DP, Konig B. Eosin Y catalyzed visible light oxidative C–C and C–P bond formation. Organic Letters. 2011;**13**:3852

[65] Kee CW, Chan KM, Wong MW, Tan C-H. Selective bromination of sp3 C–H bonds by organophotoredox catalysis. The Asian Journal of Organic Chemistry. 2014;**3**:546

[66] Ohkubo K, Mizushima K, Iwata R, Souma K, Suzuki N, Fukuzumi S. Simultaneous production of p-tolualdehyde and hydrogen peroxide in photocatalytic oxygenation of p-xylene and reduction of oxygen with 9-mesityl-10-methylacridinium ion derivatives. Chemical Communications. 2010;**46**:601

[67] Ohkubo K, Fujimoto A, Fukuzumi S. Metal-free oxygenation of cyclohexane with oxygen catalyzed by 9-mesityl-10-methylacridinium and hydrogen chloride under visible light irradiation. Chemical Communications. 2011;**47**:8515

[68] Hamilton DS, Nicewicz DA. Direct catalytic anti-markovnikov hydroetherification of alkenols. Journal of the American Chemical Society. 2012;**134**:18577

[69] Perkowski AJ, Nicewicz DA. Direct catalytic anti-markovnikov addition of carboxylic acids to alkenes. Journal of the American Chemical Society. 2013;**135**:10334

[70] Nguyen TM, Nicewicz DA. Anti-markovnikov hydroamination of alkenes catalyzed by an organic photoredox system. Journal of the American Chemical Society. 2013;**135**:9588

# 2

# Au Nanoparticle Synthesis Via Femtosecond Laser-Induced Photochemical Reduction of [AuCl$_4$]$^-$

Mallory G. John, Victoria Kathryn Meader and
Katharine Moore Tibbetts

## Abstract

Laser-assisted metallic nanoparticle synthesis is a versatile "green" method that has become a topic of active research. This chapter discusses the photochemical reaction mechanisms driving [AuCl$_4$]$^-$ reduction using femtosecond-laser irradiation, and reviews recent advances in Au nanoparticle size-control. We begin by describing the physical processes underlying the interactions between laser pulses and the condensed media, including optical breakdown and supercontinuum emission. These processes produce a highly reactive plasma containing free electrons, which reduce [AuCl$_4$]$^-$, and radical species producing H$_2$O$_2$ that cause autocatalytic growth of Au nanoparticles. Then, we discuss the reduction kinetics of [AuCl$_4$]$^-$, which follow an autocatalytic rate law in which the first- and second-order rate constants depend on free electrons and H$_2$O$_2$ availability. Finally, we explain strategies to control the size of gold nanoparticles as they are synthesized; including modifications of laser parameters and solution compositions.

**Keywords:** femtosecond laser pulses, nanocolloids, optical breakdown, gold nanoparticles, in-situ spectroscopy, photochemical reduction mechanisms

## 1. Introduction

The unique chemical and physical qualities of metallic nanoparticles have attracted the attention of researchers. Their size- and shape-dependent optical properties make them especially appealing due to the potential technological applications [1–4]. In particular, gold nanoparticles (AuNPs) have strong absorptions in the visible spectrum that come from the collective oscillations of surface conduction-band electrons as they interact with light, which is called the surface plasmon resonance (SPR). The dependence of the SPR absorption on particle size and shape opens a range of application possibilities for AuNPs, including surface enhanced Raman spectroscopy [5]; non-invasive diagnostic imaging [2]; photothermal cancer therapy [3, 6]; plasmon-enabled photochemistry; and catalytic reactions such as water-splitting [7, 8]. It is necessary to these ends that the NP sizes and shapes are controllable during synthesis [9, 10]. Control can be achieved chemically, by modifying experimental conditions like temperature, reaction time, metal-ion concentration, and the absence or presence of reducing agents and surfactants [2]. Laser-assisted approaches to AuNP fabrication allow the manufacture of "pure" NPs which lack chemical reducing agents or surfactants, making this synthesis method ideal for NPs intended to be used in catalysis, and other electronic, biological or medical applications [11, 12].

There are two common approaches to colloidal AuNP synthesis using laser-assisted methods. The first is bulk-metal ablation, in which metal atoms are ejected from the target material and form nanoparticles in solution [13]. The second is by irradiating a metal-salt solution to produce reducing agents via solvent-molecule photolysis [14–16]. Controlling nucleation and growth of the nanoparticles during metal-salt reduction by changing laser parameters (focusing conditions, pulse duration, pulse energy, irradiation time), and chemical parameters (metal-ion concentration, solvent composition, presence of capping agents), determines the size, shape, and stability of the colloidal products [14–30]. Laser-assisted AuNP fabrication requires a simple setup, which facilitates experimentation [12].

Section 2 of this chapter examines $[AuCl_4]^-$ reduction under femtosecond-laser irradiation, specifically the microplasma formation that arises from optical breakdown (OB) and super continuum emission (SCE). We review both theoretical models for OB and SCE, and provide experimental measurements of both OB and SCE to show which dominates each set of experimental conditions. In Section 3, we describe the chemical reactions that cause photochemical $[AuCl_4]^-$ reduction and AuNP formation, and compare them with the observed autocatalytic reduction kinetics. We relate observed first- and second-order autocatalytic rate constants to the availability of reducing species in the microplasma. Lastly, in Section 4, we review recent literature that describes control over AuNP size- and shape-control through manipulation of laser conditions and chemical composition of the solution.

## 2. Background: interactions of ultrashort laser pulses with condensed media

In a dielectric medium with a band gap that exceeds the laser photon-energy, ultrashort laser pulses can produce quasi-free electrons in the conduction band by two processes: (1) nonlinear multiphoton ionization and tunneling photoionization [31], and (2) high-kinetic-energy free electron collisions with neutral molecules, causing cascade ionization, also called avalanche ionization [32]. The formation of free electrons initially generates a localized, weakly-ionized plasma [32–34], which can initiate optical breakdown (OB), supercontinuum emission (SCE), or both [35–37]. This section provides an overview of the theory behind both processes, and some experimental measurements.

## 2.1. Optical breakdown

Optical breakdown (OB) of a transparent dielectric medium occurs when the free-electron density $\rho_e$ in plasma exceeds a critical value, and depends on the peak intensity $I$ of the excitation pulse [32–34]. Recent experiments in water have quantified the critical value for $\rho_e$ as the threshold for cavitation-bubble formation at $\rho_e = 1.8 \times 10^{20}$ cm$^{-3}$ [38]. In order to calculate the electron density resulting from the laser–medium interaction, media such as water and other solvents are typically modeled as a dielectric, with band gap $\Delta$. For water, the band gap is usually specified as $\Delta = 6.5$ eV [33, 34], although some recent experiments have placed the effective band gap as high as 9.5 eV for direct excitation into the conduction band [39, 40].

Conventionally, the laser pulse propagates in the $z$ direction with a time-dependent Gaussian intensity envelope based on the focusing conditions [41],

$$I(z,t) = \frac{P(t,z)}{A(z)} = \frac{E_p}{\tau_p \pi w(z)^2} \exp\left[ (-4\ln 2)\left(\frac{t - z/c}{\tau_p}\right)^2 \right]; \quad w(z) = w_0 \sqrt{1 + \frac{z^2}{z_R^2}} \qquad (1)$$

where $P(t,z)$ is the time-dependent power density, $A(z)$ the cross-sectional area, $E_p$ the pulse energy, $\tau_p$ the pulse duration, $c$ the speed of light, $w_0$ the beam waist at the focus, and $z_R$ the Rayleigh range.

The time evolution of the free-electron density $\rho_e$ produced by the laser–water interaction is governed by the differential equation [33]

$$\frac{\partial \rho_e}{\partial t} = W_{\text{photo}} + W_{\text{casc}}\rho_e - W_{\text{diff}}\rho_e - W_{\text{rec}}\rho_e^2. \qquad (2)$$

Free electrons are produced according to the photoionization rate $W_{\text{photo}}$ and cascade ionization rate $W_{\text{casc}}$, while electrons are lost from the focal volume at diffusion rate $W_{\text{diff}}$, and recombination rate $W_{\text{rec}}$. The specific formulas describing each rate are reviewed elsewhere [33, 34].

At a given laser wavelength, the peak-intensity needed to reach critical electron density for OB is highly dependent on pulse duration, due to the interplay between the photoionization and cascade ionization rates [33, 34]. To illustrate this effect over a wide range of pulse durations used in recent [AuCl$_4$]$^-$ photochemical reduction studies [14–30], Eq.(2) coupled to the appropriate formulas for each rate [33, 34] was solved using the Runge–Kutta integrator ode45 incorporated into MATLAB, as in our previous work [16]. The critical electron density threshold was taken to be the recently reported experimental value $\rho_e = 1.8 \times 10^{20}$ cm$^{-3}$ [38]. **Figure 1(a)** shows the calculated time-dependent electron density $\rho_e(t)$ for pulses at intensity $I = 10^{13}$ W cm$^{-2}$ with durations of 30 fs (dark blue), 100 fs (light blue), 200 fs (green), 1.5 ps (orange), and 36 ps (red). The value of zero on the abscissa corresponds to the center of the pulse, and the time is normalized to the respective pulse durations. The dashed line at $\rho_e = 1.8 \times 10^{20}$ cm$^{-3}$ indicates the OB threshold. The rise in peak electron density with pulse duration results from the increased

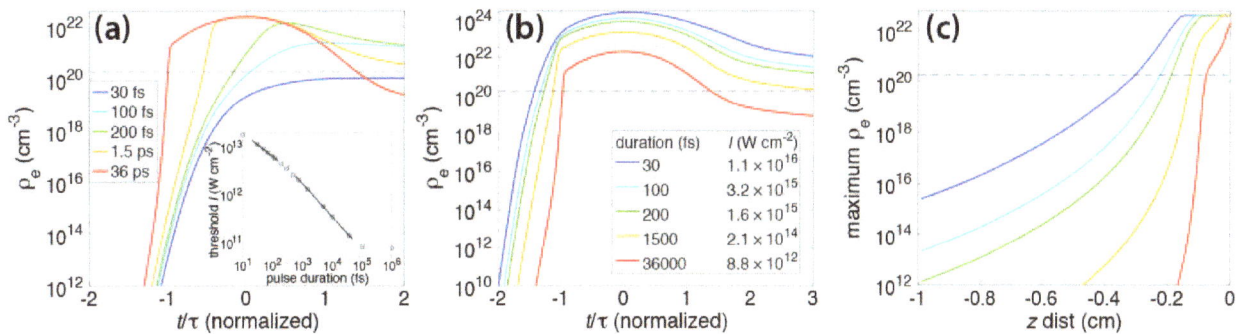

**Figure 1.** (a) Electron density vs. time for $1 \times 10^{13}$ W cm$^{-2}$ pulses with a series of durations from 30 fs to 36 ps. Inset: Threshold intensity required to achieve OB as a function of pulse duration. (b) Electron density vs. time for 1 mJ pulses. (c) Electron density vs. propagation distance $z$ from the geometric focus for 1 mJ pulses.

contribution of cascade ionization to the formation of free electrons as the pulse lengthens [33, 34]. As a result, the threshold intensity to achieve OB decreases by two orders of magnitude as the pulse duration is increased from 30 fs to 36 ps (inset, **Figure 1(a)**).

The high pulse-energies of up to 5 mJ and tight-focusing conditions often used in [AuCl$_4$]$^-$ reduction experiments [14–21] produce peak intensities that significantly exceed the OB threshold. For instance, irradiation with 1 mJ pulses under the conditions described above results in a peak electron-density that surpasses the OB threshold by at least factor of 50 and even exceeds the maximum electron-density of $4 \times 10^{22}$ cm$^{-3}$ achievable in liquid water [42] for shorter pulses (dotted line, **Figure 1(b)**). Thus, to model the availability of electrons for [AuCl$_4$]$^-$ reduction, it is of primary importance to estimate the plasma volume in which the electron-density exceeds the OB threshold. The plasma volume may be estimated by calculating the critical distance $z_{crit}$ in front of the focus where the OB threshold is exceeded. The value of $z_{crit}$ for a given pulse energy, duration, and focusing-geometry may be calculated by solving Eq. (2) for a Gaussian beam in Eq. (1) at a series of propagation distances $z < 0$ cm (i.e., before the focal spot at $z = 0$ cm) in order to determine the highest electron density achieved. The resulting peak electron-density as a function of $z$ is shown for the series of 1 mJ pulses from **Figure 1(b)** in **Figure 1(c)**. As the pulse duration decreases, OB begins farther from the focus, with $z_{crit}$ increasing from 0.1 cm for 36 ps pulses to 0.3 cm for 30 fs pulses. This result shows that the plasma volume, which is proportional to $z_{crit}^3$, depends strongly on both pulse energy and duration in a given experiment. Our earlier simulations have shown that for a series of pulse durations with the same focusing-geometry, $z_{crit}$ grows with peak-intensity as $z_{crit} \propto I^{1/2}$, meaning that the plasma volume grows as $I^{3/2}$ [16]. As discussed below, the growth of plasma volume is directly proportional to the [AuCl$_4$]$^-$ reduction rate.

## 2.2. Supercontinuum emission and filamentation

The filamentation process leading to SCE arises from self-focusing of the laser pulse in a nonlinear Kerr medium. A full discussion of the details of nonlinear light propagation leading self-focusing is beyond the scope of this work and may be found in Refs. [35–37]. Briefly, filamentation depends on the laser power $P$ and is initiated when $P$ exceeds the critical power $P_{crit}$ [37, 43]

$$P_{\text{crit}} = \frac{3.77\lambda^2}{8\pi n_0 n_2} \tag{3}$$

where $\lambda$ is the laser's wavelength, $n_0$ is the refractive index of the medium, and $n_2$ characterizes the intensity-dependent refractive index $n = n_0 + n_2 I$. In water, $P_{\text{crit}}$ has been measured at $4.2 \times 10^6$ W for 800 nm pulses [44], which translates into very modest pulse energies of 0.13 and 0.42 $\mu$J at 30 and 100 fs. Filamentation causes spectral broadening to both the red and blue of the laser wavelength. A red-shift is caused by rotational and vibrational motion of the molecules in the medium, and a blue-shift happens when the power $P$ is high enough to form a shockwave at the trailing temporal edge of each pulse [36]. Blue-shifts produce a broad pedestal as far as 400 nm in the output-spectrum for pulses shorter than 100 fs [16, 27, 28, 35–37, 45] (see also **Figure 2**). Because SCE depends on power instead of peak intensity, filamentation may occur at intensities below the OB threshold; especially when the laser beam is weakly-focused or collimated [35–37, 45–49]. For laser beams with peak-intensities on the order of $10^{12}$ W cm$^{-2}$, the filament electron-density has been measured at $1 - 3 \times 10^{18}$ cm$^{-3}$ [48, 49]. Such weakly-ionized SCE plasmas can drive [AuCl$_4$]$^-$ reduction even in the absence of OB [26–28], while the white light from the SCE has been shown to induce AuNP-fragmentation by resonant absorption and Coulomb explosion [45–47].

## 2.3. Experimental measurement of OB and SCE

The presence of OB and SCE may be measured with a spectrometer arranged as shown in **Figure 2(a)** [35, 36]. To detect OB from light that has scattered off of the OB plasma, the fiber mount is placed at a 90° angle to the laser beam (geometry (i) in **Figure 2(a)**) and a series of lenses focuses the light into the fiber mount. For SCE detection, the fiber mount is placed along the beam path, behind the sample (geometry (ii) in **Figure 2(a)**). A diffuser is attached to the fiber mount to avoid saturating the spectrometer. For tightly focused beams, OB is expected at any pulse duration, and SCE may also be present if the pulse is short. When the beam is loosely-focused or collimated, only SCE is expected.

To illustrate the conditions in which OB, SCE, or both are present, tightly focused pulses [16] and unfocused, collimated pulses [28] were measured using the setup in **Figure 2(a)**. **Figure 2(b)** shows spectra obtained at detector (i) for tightly focused 30 and 1500 fs pulses at 0.3 and 0.03 mJ pulse energy. The broadened spectrum for 30 fs pulses indicates the presence of SCE

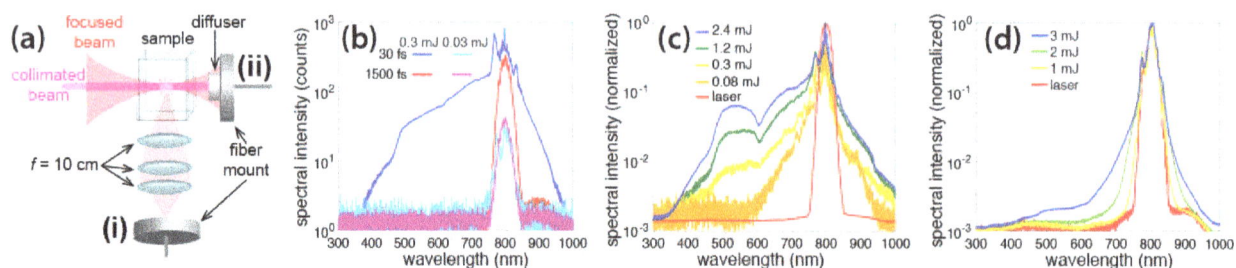

**Figure 2.** (a) Setup for OB and SCE measurements. (b) OB spectra for tightly focused 30 fs and 1500 fs pulses. (c) SCE spectra for tightly focused 30 fs pulses. (d) SCE spectra for collimated 30 fs pulses.

along with OB, while the narrow spectrum with 1500 fs pulses indicates that no SCE occurs. **Figure 2(c)** and **(d)** show SCE spectra obtained for 30 fs pulses at a series of pulse energies for tightly focused and collimated pulses. Under both conditions, asymmetric broadening towards the visible region of the spectrum grows with increasing pulse energy. The spectral broadening saturates for tightly focused pulses at energies above 1.2 mJ (**Figure 2(c)**), while greater pulse-energies would be needed to saturate the spectral width for unfocused pulses (**Figure 2(d)**). No OB is observed when the beam is collimated, indicating that a low-density plasma (LDP) with $\rho_e \sim 10^{18}$ cm$^{-3}$ is present in the filaments [48, 49]. LDP conditions have been used by research groups to control the synthesis of AuNPs [26, 28, 45, 50].

# 3. Mechanisms of [AuCl$_4$]$^-$ reduction

## 3.1. Reactions of water

The key role that water photolysis plays, in the photochemical reduction of [AuCl$_4$]$^-$ and other metal salts, is well-established [14–22, 26, 27], and supported by the presence of H$_2$, O$_2$, and H$_2$O$_2$ as water is irradiated with high-intensity femtosecond laser pulses [14, 19, 51]. Two common mechanisms proposed to explain the reduction of aqueous [AuCl$_4$]$^-$ under high-intensity laser irradiation are (a) direct homolysis of the Au-Cl bond by multiphoton absorption to form Au(II) and Au(I) intermediates, and (b) chemical reduction of Au(III) ions by the reactive species formed from water photolysis [15–19, 26]. Since the number of water molecules far surpasses the number of [AuCl$_4$]$^-$ molecules in solution, the second proposed mechanism is more likely for [AuCl$_4$]$^-$ reduction to Au(0) in aqueous solutions. The photolysis reactions involved include [27, 52–55]

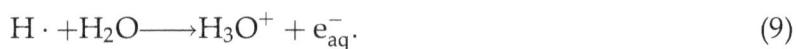

$$H_2O \xrightarrow{nh\nu} e^- + H^+ + OH\cdot \tag{4}$$

$$e^- \longrightarrow e_{aq}^- \tag{5}$$

$$e_{aq}^- + OH\cdot \longrightarrow OH^- \tag{6}$$

$$H_2O \xrightarrow{nh\nu} H\cdot + OH\cdot \tag{7}$$

$$2OH\cdot \longrightarrow H_2O_2 \tag{8}$$

$$H\cdot + H_2O \longrightarrow H_3O^+ + e_{aq}^-. \tag{9}$$

Although both hydrated-electrons and hydrogen radicals are capable of reducing [AuCl$_4$]$^-$, the fast consumption of H. via Eq. (9) observed in water photolysis using picosecond pulses [53] suggests an inconsequential contribution by H. to [AuCl$_4$]$^-$ reduction. In contrast, hydrated electrons may be formed from both the free electrons generated in OB plasma via Eq. (5) within several hundred femtoseconds [54, 55], and from the reaction of water with H. via Eq. (9). Hydrated electrons have lifetimes of up to hundreds of nanoseconds in pure water [56] and react with $[AuCl_4]^-$ with a diffusion-controlled rate constant of $6.1 \times 10^{10}$ $M^{-1}$ $s^{-1}$ [57]. There-fore, hydrated electrons are the dominant $[AuCl_4]^-$ reducing agent through the reaction [27]

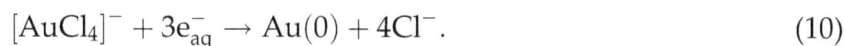

$$[AuCl_4]^- + 3e_{aq}^- \rightarrow Au(0) + 4Cl^-. \tag{10}$$

Another product of water photolysis, $H_2O_2$, is generated from the recombination of two hydroxyl radicals via Eq. (8) and drives $[AuCl_4]^-$ reduction and AuNP formation [15, 16, 19, 20]. Tangeysh et al. explored the role that $H_2O_2$ played in $[AuCl_4]^-$ reduction by monitoring the UV–vis absorbance of $[AuCl_4]^-$ samples after laser-irradiation termination, but before all of the $[AuCl_4]^-$ had been consumed [19]. They explained the post-irradiation $[AuCl_4]^-$ reduction and SPR absorbance-peak growth by proposing that the $H_2O_2$ produced during irradiation reduced the remaining $[AuCl_4]^-$, in the presence of the existing AuNPs [19]. This hypothesis was developed further by Tibbetts et al. [15], using previous work showing that $H_2O_2$ reduces $[AuCl_4]^-$ in the presence of AuNPs via the reaction [58, 59]

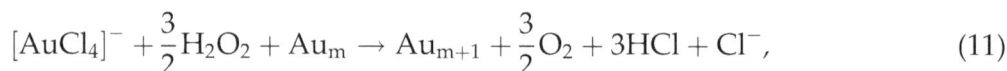

$$[AuCl_4]^- + \frac{3}{2}H_2O_2 + Au_m \rightarrow Au_{m+1} + \frac{3}{2}O_2 + 3HCl + Cl^-, \tag{11}$$

where the existing AuNPs act as a catalyst for $[AuCl_4]^-$ reduction. This process underlies the observed autocatalytic reduction kinetics of $[AuCl_4]^-$.

## 3.2. Kinetics

Controlling the sizes and shapes of AuNPs starts with kinetic control of their nucleation and growth. LaMer's nucleation theory, developed in 1950 [60], was used to describe AuNP formation first [4, 9, 61], but Turkevich's studies [62] on reduction of $HAuCl_4$ using sodium citrate yielded more appropriate AuNP formation-mechanisms, including autocatalysis [62–64] and aggregative growth [65, 66]. In 1997, Watzky and Finke described the reduction of transition metal salts using $H_2$, undergoing slow, continuous nucleation accompanied by fast, autocatalytic surface growth to form nanoparticles. They described this mechanism using a quantitative, two-step rate law [67],

$$-\frac{d[A]}{dt} = \frac{d[B]}{dt} = k_1[A] + k_2[A][B] \tag{12}$$

where [A] is the precursor (metal salt) concentration, [B] is the metal nanoparticle concentration, $k_1$ is the rate constant of metal-cluster nucleation (slow) and $k_2$ is the rate constant of autocatalytic growth of the nanoparticles (fast) [67, 68]. Integration of Eq. (12) gives the time-dependent precursor and metal nanoparticle concentrations [A($t$)] and [B($t$)] [67]

$$[A(t)] = \frac{\frac{k_1}{k_2} + [A(0)]}{1 + \frac{k_1}{k_2[A(0)]}e^{(k_1 + k_2[A(0)])t}} \tag{13}$$

$$[B(t)] = 1 - \frac{\frac{k_1}{k_2} + [A(0)]}{1 + \frac{k_1}{k_2[A(0)]}e^{(k_1 + k_2[A(0)])t}} \tag{14}$$

where [A(0)] is the initial precursor concentration. The rate law in Eq. (13) has been used to describe AuNP formation from reducing ionic precursors via wet chemical routes [67–69] and for femtosecond laser-induced $[AuCl_4]^-$ reduction under a variety of laser conditions and solution compositions [15, 16, 26, 28]. Eq. (14) follows if it is assumed that the conversion of Au(III) to Au(0) is fast enough that no significant concentration of intermediate species like Au(I) builds up.

The time-dependent concentrations of $[AuCl_4]^-$ and AuNPs needed to determine the reaction kinetics may be obtained from *in situ* UV–vis spectra recorded during laser irradiation [15]. **Figure 3(a)** displays representative absorbance spectra of $[AuCl_4]^-$ after different irradiation times. The arrow labeled 250 nm corresponds to the decrease in the LMCT band of $[AuCl_4]^-$, while the arrow labeled 450 nm corresponds to the growth of AuNPs [15, 16, 70]. To obtain the time-dependent $[AuCl_4]^-$ concentration in Eq. (13), the absorbance of $[AuCl_4]^-$ is monitored at $\lambda = 250$ nm. Because AuNPs also absorb across the UV range, the absorbance at 250 nm corresponds to the absorbance contributions from both the $[AuCl_4]^-$ precursor and AuNP product species. The $[AuCl_4]^-$ contribution can be isolated from the 250 nm absorbance by subtracting off the AuNP contribution, as described in previous work [15, 16]. Alternatively, monitoring the absorbance at $\lambda = 450$ nm where only the AuNPs absorb [70] allows direct monitoring of the - time-dependent AuNP growth. Both representations of the reaction kinetics are shown in **Figure 3**: (b) normalized absorbance of $[AuCl_4]^-$ at 250 nm and (c) 450 nm as a function of laser irradiation time for focused 30 fs laser pulses at a series of pulse energies [16]. The dots denote the experimental data, and the solid lines are fits to Eq. (13) (**Figure 3(b)**) or Eq. (14) (**Figure 3(c)**). The disappearance rate of $[AuCl_4]^-$ and growth rate of AuNPs mirror each other, showing that the rate constants may be extracted from fitting either spectral absorbance. In practice, small amounts of intermediate species such as Au(I) are present during photochemical reduction [15], so the rate constants extracted from fitting the normalized 450 nm absorbance to Eq. (14) are $20 - 50\%$ lower than those from fitting the normalized 250 nm absorbance to Eq. (13).

Under certain experimental conditions where the $[AuCl_4]^-$ reduction rate is slowed, the experimental kinetics are more accurately modeled by adding a linear component to Eq. (13) [15, 28],

$$[A(t)] = \frac{\frac{k_1}{k_2} + [A(0)]}{1 + \frac{k_1}{k_2[A(0)]} e^{(k_1+k_2[A(0)])t}} - k_3 t \tag{15}$$

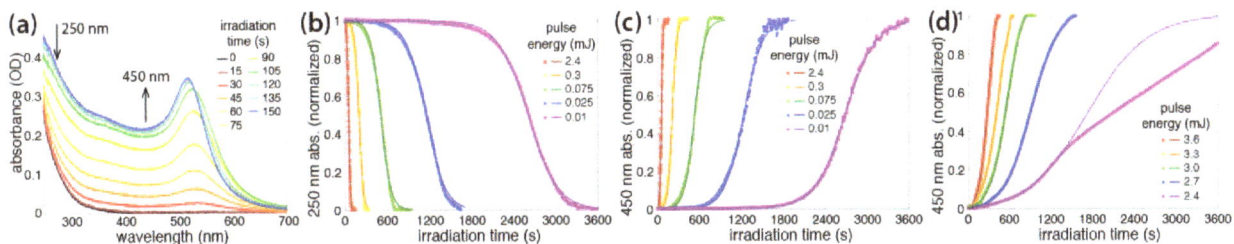

**Figure 3.** (a) UV–vis spectra of $[AuCl_4]^-$ solution irradiated for different times. Representative plots of normalized absorbance at 250 nm (b) and 450 nm (c), (d) (dots) as a function of irradiation time for the pulse energies labeled in the legend, with fits to Eq. (13) (b), Eq. (14) (c), and Eq. (16) (d) (solid lines). Data taken from refs. [16, 28].

$$[B(t)] = 1 - \frac{\frac{k_1}{k_2} + [A(0)]}{1 + \frac{k_1}{k_2[A(0)]} e^{(k_1 + k_2[A(0)])t}} + k_3 t. \tag{16}$$

The third rate constant, $k_3$, is zeroth order with respect to the $[AuCl_4]^-$ concentration, and was suggested arise due to limited availability of reducing species from photolysis of the solvent [15]. This rate equation more accurately fits the reduction kinetics when the laser beam is collimated such that LDP conditions are present [28], as shown in **Figure 3(d)**. When the pulse energy is sufficiently low (2.4 mJ), the AuNP growth was significantly slower due to agglomeration of the formed nanoparticles; therefore, only the first portion of the experimental data was fit to Eq. (16). Similar agglomeration has been observed in other experiments conducted under LDP conditions, in which the initial portion of experimental data was fit to Finke-Watsky kinetics [26].

Extracting the rate constants at a series of experimental conditions (e.g., solution pH [15], pulse energy [16, 28]) provides significant insight into the roles that the reactions in Eqs. (4)–(9) play in the conversion of $[AuCl_4]^-$ to AuNPs. **Figure 4(a)** and **(b)** show the rate constants $k_1$ and $k_2$ extracted from fits to Eqs. (13) and (16), respectively, for tightly focused and collimated 30 fs pulses, as a function of pulse energy (proportional to peak intensity $I$) [16, 28]. In the log–log plots, the slopes of the least squares fit lines denote the power law dependence of each rate constant on the peak intensity. For the tight focusing geometry in **Figure 4(a)**, the nucleation rate constant grows as $k_1 \sim I^{(1.6\pm0.2)}$, a factor of three faster than the growth of the autocatalytic rate constant $k_2 \sim I^{(0.56\pm0.02)}$. In contrast, under LDP conditions, $k_1$, $k_2$, and $k_3$ all grow approximately as $I^4$ (**Figure 4(b)**).

The power law dependence of $k_1$ under tight-focusing conditions corresponds to the growth of the OB plasma volume $V$ with peak-intensity $I$ as $V \sim I^{1.5}$ calculated via Eq. (2) (c.f., Section 2.1) [16]. **Figure 4(c)** shows that $k_1 \sim V$ for pulse durations ranging from 30 to 1500 fs under tight-focusing conditions. This result demonstrates the that the production of hydrated electrons an OB plasma controls the rate of nucleation of $[AuCl_4]^-$. The higher sensitivity of $k_1 \sim I^4$ under LDP conditions [28] is consistent with the electron density in LDP conditions being proportional to the multiphoton order required for ionization of the medium [37], where 5 photons at 800 nm are needed to ionize water [34]. The importance of hydrated electrons to

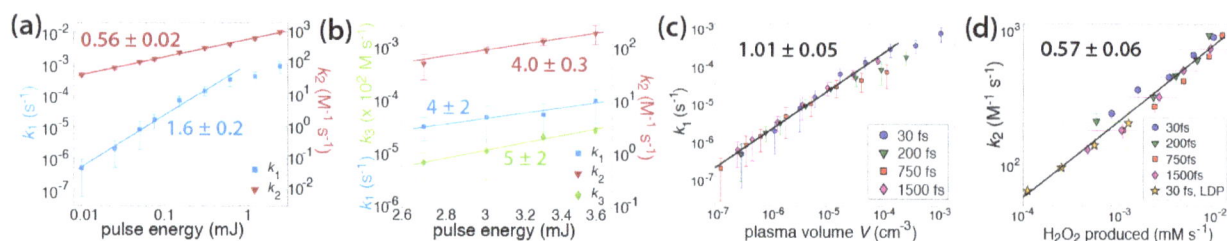

**Figure 4.** Rate constants for 30 fs laser pulses (blue, $k_1$; red, $k_2$; green $k_3$) as a function of pulse energy for (a) tight focusing geometry and (b) LDP geometry. (c) Correlation between $k_1$ rate constant and calculated OB plasma volume for tight focusing geometry. (d) Correlation between $k_2$ rate constant with $H_2O_2$ production for tight focusing and LDP geometries. Data taken from refs. [16, 28].

$[AuCl_4]^-$ reduction has been demonstrated in two other works [15, 26]. The addition of $N_2O$ as a hydrated electron scavenger to aqueous $[AuCl_4]^-$ significantly lowers the value of $k_1$ in Eq. (14) with respect to $k_2$, thereby isolating the contribution of hydrated electrons to $[AuCl_4]^-$ nucleation [26]. Increasing the solution pH of aqueous $[AuCl_4]^-$ through addition of KOH significantly increases $k_1$ [15], which is consistent with previous findings that the lifetimes of hydrated electrons are suppressed in acidic solution [55].

Under both tight-focusing and LDP conditions, the power law dependence of $k_2$ corresponds to the formation rate of $H_2O_2$ from water. Under tight-focusing conditions, the formation rate of $H_2O_2$ is $[H_2O_2] \sim I$, so the lower dependence of $k_2 \sim I^{1/2}$ results in the relationship $k_2 \sim [H_2O_2]^{1/2}$ [16]. The same correlation was found under LDP conditions, where the formation rate of $H_2O_2$ is $[H_2O_2] \sim I^8$ and $k_2 \sim I^4$ [28]. Both correlations between $[H_2O_2]$ and $k_2$ are shown in **Figure 4(d)**. These results are consistent with other work in which the addition of the OH. scavenger 2-propanol [26] resulted in increased $k_1$ values relative to $k_2$ values and slower growth of AuNPs.

The kinetics results quantifying the dependence of both nucleation and autocatalytic growth rate constants demonstrate the importance of both short and long-lived reducing species to control the formation of AuNPs via photochemical reduction of aqueous $[AuCl_4]^-$. The reactive species produced during water photolysis can be controlled by changing both the laser irradiation conditions [16, 28] and the chemical composition of the $[AuCl_4]^-$ solution [15, 26]. The following section will review how changing both of these reaction conditions can control the size and shape of the synthesized AuNPs.

# 4. Controlling Au nanoparticle sizes

Several recent articles have reported some degree of control over the size of AuNPs synthesized by photochemical reduction of $[AuCl_4]^-$ through the manipulation of experimental conditions: broadly, the laser parameters and solution composition. The focusing-geometry, pulse energy, and pulse duration determine the generation of OB and SCE, which direct AuNP growth [16, 20, 28]. Adding scavengers and modifying the solution pH also change the $[AuCl_4]^-$ reduction kinetics, and therefore, particle size [15, 26]. Finally, adding capping can produce smaller AuNPs [14, 19, 20, 29, 30].

## 4.1. Laser parameters

### 4.1.1. Focusing-geometry

Focusing-geometries influence the nature of the nonlinear interactions between the laser and solution (c.f., Section 2). Without strong-focusing, SCE yields a LDP environment containing electron-densities on the order of $\rho_e \sim 10^{18}$ cm$^{-3}$. This setup has been used for photochemical $[AuCl_4]^-$ reduction [26, 28] experiments and Au ablation [45, 50] experiments. LDP conditions seem well-suited to applications like AuNP synthesis, because second-order reactions are suppressed, including those that yield $H_2O_2$ [26, 50]. Without abundant reducing species, $[AuCl_4]^-$ to AuNP conversion is slow. Many research groups opt to use a focused-geometry [14–20], which yield electron densities that exceed the OB threshold, $\sim 10^{20}$ cm$^{-3}$. Low numerical aperture (NA) geometries produce SCE through self-focusing and filamentation processes. These processes can cause intensity-clamping, which stops the intensity from exceeding $I \sim 10^{13}$ W cm$^{-2}$ [37], and limits the number of reactive species available for reduction [20]. In contrast, tight-focusing (high-NA) geometries, simultaneous spatial and temporal focusing (SSTF) [71], or spatial beam-shaping [72] can avoid excessive filamentation and intensity clamping. In Au nanoparticle synthesis, tight-focusing [14, 16, 18] and SSTF [15, 19, 20], where the frequency components of the laser pulse are spatially separated prior to focusing, have both been used for this purpose. **Figure 5** shows schematic diagrams (top) and photographs (bottom) of fs-laser irradiation of water using (a) collimated beam geometry [28], (b) low-NA focusing [20], (c) high-NA focusing [16], and (d) SSTF [19]. The absence of visible filaments in panels (c) and (d), compared to (b), suggest less intensity-clamping, and so a higher peak-intensity at the focal spot.

Another phenomenon to consider when experimenting with focusing-conditions is cavitation bubble formation, which happens when the OB electron-density threshold is exceeded [38]. The generated cavitation bubbles are sensitive to the shape of the laser-plasma [20, 72]. Under low-NA focusing-conditions with filamentation and SCE, the bubbles are ejected from the focus with low kinetic energy, seen as a small stream of bubbles rising from the center of the cuvette in **Figure 5(b)**. This condition results in inefficient and asymmetrical mixing-dynamics of the reactive species throughout the solution, but can be improved with the addition of a magnetic stir-bar [20]. Similarly, a stir bar is needed when operating under LDP conditions to ensure that the solution is being mixed [26, 28]. Both high-NA focusing and SSTF produce more spherical plasmas, which eject high kinetic energy bubbles into solution radially [72], causing turbulent mixing of the reactive species into the solution (evident in **Figure 5(c)**) and removing the need for stirring [16, 20].

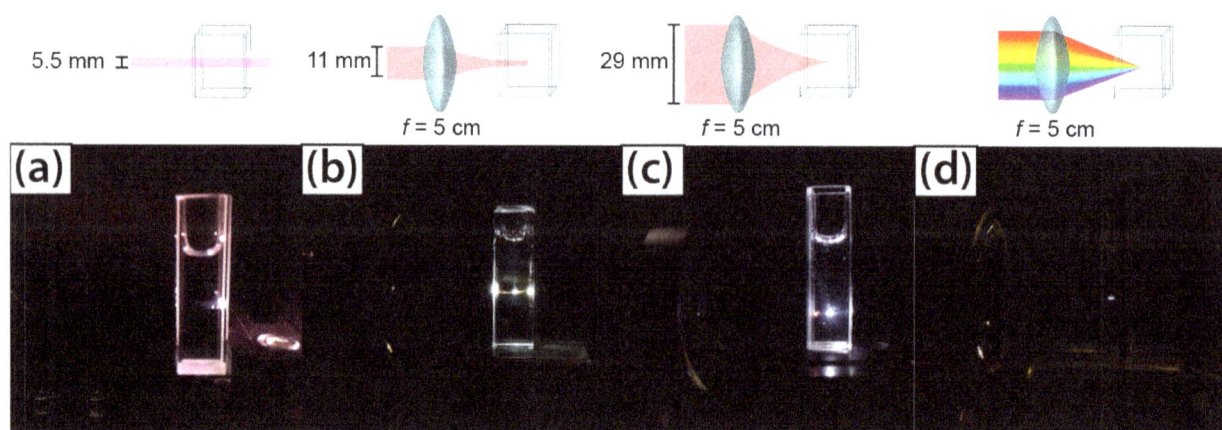

**Figure 5.** Diagrams (top) and photographs (bottom) of irradiated water using (a) collimated beam, (b) low-NA focusing, (c) high-NA focusing, and (d) SSTF.

| Ref. | Condition | Energy (mJ) | Size (nm) |
|---|---|---|---|
| [26] | LDP | 1.35 | $29.1 \pm 17.3$ |
| [28] | LDP | 2.7 | $27.1 \pm 7.0$ |
| [20] | Low-NA | 1.8 | $13.6 \pm 8.0$ |
| [18] | High-NA | 5.6 | $4.0 \pm 1.7$ |
| [16] | High-NA | 2.4 | $3.5 \pm 1.9$ |
| [20] | SSTF | 1.8 | $10.2 \pm 4.1$ |
| [15] | SSTF | 2.5 | $9.2 \pm 4.1$ |

**Table 1.** Reported laser focusing conditions and resulting AuNP sizes.

The complex interactions between high-intensity fs laser pulses and aqueous solution result in the particle size being sensitive to different focusing-conditions. **Table 1** summarizes the results of AuNP syntheses prepared in aqueous solutions without capping agents across focusing conditions. Of the reported sizes, the largest AuNPs resulted from the LDP conditions [26, 28], which limits production of the reducing species, driving AuNP formation through aggregative growth and agglomeration. Smaller AuNPs were formed with the high-NA focusing conditions [16, 18], which produces high electron-densities because of tight laser-focus; filamentation is suppressed, and the reducing species (electrons and $H_2O_2$) are thoroughly mixed throughout the solution by the OB plasma. In combination, these conditions generate many Au(0) seeds but limit Au(III) ions. SSTF and low-NA focusing-geometries create intermediate AuNP sizes. The SSTF focusing-geometry improves size-distribution because it mixes the reactive species well with its spherical plasma [20]. Adjusting the laser's focus-geometry has a strong influence on both the production rate and spatial distribution of the reducing species required for AuNP formation, and yields another dimension of control over particle sizes.

### 4.1.2. Pulse energy and duration

Several studies on focusing-conditions have demonstrated that increasing the pulse energy reduces AuNP size [16, 20, 28]. When tight-focusing geometry was used, increasing the energy of a 30 fs pulse from 0.15 mJ to 2.4 mJ decreased AuNP size from $6.4 \pm 5.6$ nm to $3.5 \pm 1.9$ nm [16] (**Figure 6(a)** and **(b)**). This trend was also seen when LDP conditions were used: increasing the energy of 30 fs pulses from 2.7 to 3.3 mJ reduced AuNP size from $27 \pm 7$ to $14 \pm 6$ nm [28] (**Figure 6(c)** and **(d)**). These results are consistent with earlier reports using SSTF with 36 ps pulses to irradiate solutions of $[AuCl_4]^-$ and polyethylene glycol (PEG), a capping agent. Increasing the pulse energy from 0.45 mJ to 1.8 mJ reduced the average particle size from $9.6 \pm 2.7$ to $5.8 \pm 1.1$ nm [20].

While the pulse energy strongly influences the size of the AuNPs from photochemical reduction of $[AuCl_4]^-$, the pulse duration, or linear frequency chirp, has at most a modest effect on the AuNP size at a fixed pulse energy and focusing condition [16, 20]. Under tight focusing conditions, stretching the pulse duration was stretched from 30 to 1500 fs (negatively chirped) at a 0.15 mJ pulse energy slightly decreased the AuNP sizes from $6.4 \pm 5.6$ to $4.4 \pm 4.0$ nm [16].

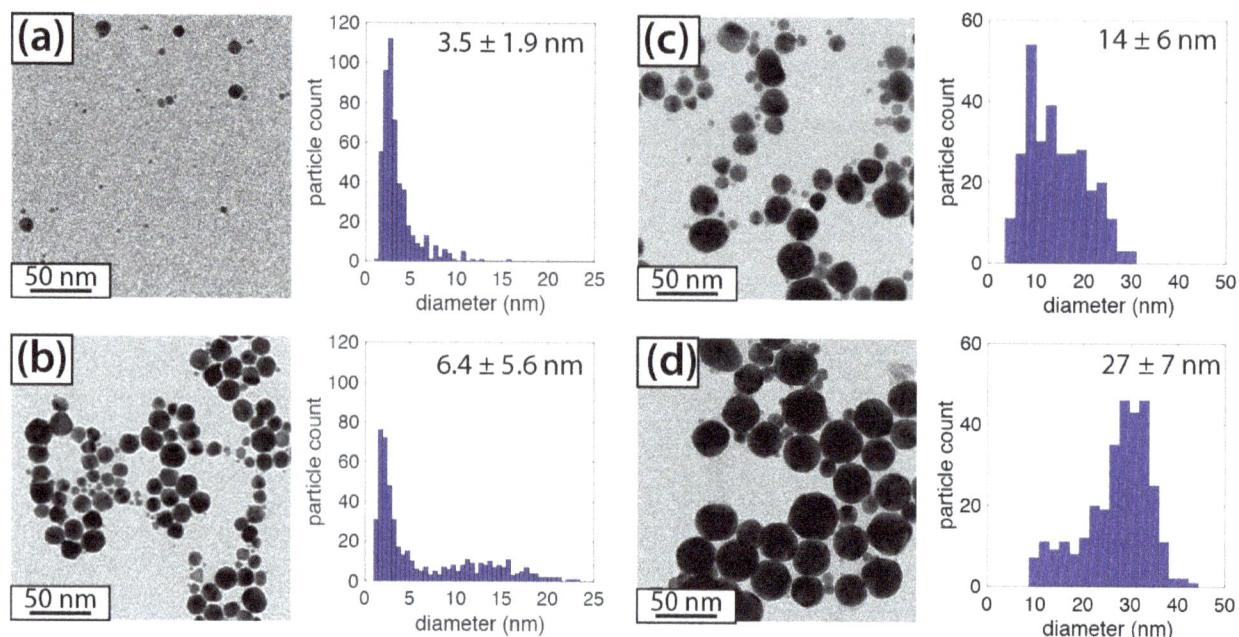

**Figure 6.** Representative TEM images and AuNP size distributions synthesized with 30 fs pulses under the following conditions: (a) tightly focused, 2.4 mJ; (b) tightly focused, 0.15 mJ; (c) LDP, 3.3 mJ; and (d) LDP, 2.7 mJ.

When the experiment was repeated at a high pulse energy (2.4 mJ), the AuNP size increased from $3.5 \pm 1.9$ nm for 30 fs pulses to $6.3 \pm 2.4$ nm for 1500 fs pulses [16]. In a separate experiment using low-NA focusing conditions, 1.8 mJ pulses with chirp coefficients of $+20,000$ $fs^2$, $0$ $fs^2$, and $-20,000$ $fs^2$ (corresponding to 35 fs unchirped pulses and 2 ps chirped pulses) produced $8.2 \pm 3.5$, $8.1 \pm 3.4$, and $8.1 \pm 6.5$ nm AuNPs, respectively [20]. Collectively, these results suggest that that for sufficiently high peak intensities generating OB conditions, the pulse duration does not significantly affect the size of AuNPs produced by photochemical reduction of $[AuCl_4]^-$.

## 4.2. Chemical composition

### 4.2.1. Scavengers

Water photolysis produces reactive species, which govern $[AuCl_4]^-$ reduction and therefore AuNP formation. To manage particle growth, radical scavengers can be added to solution. As summarized in Section 3, it is primarily the hydrated electrons that reduce $[AuCl_4]^-$ (Eq. (10)). $H_2O_2$ (formed by the recombination of two hydroxyl radicals (Eq. (8)), facilitates autocatalytic particle growth (Eq. (11)). Scavengers can selectively hinder the production of water photolysis byproducts such as $H_2O_2$ [73], so they have been used to control reduction kinetics [26].

The hydrated-electron scavenger $N_2O$, and hydroxyl radical scavengers 2-propanol and ammonia, were originally studied in water radiolysis using X-rays and $\gamma$ rays [74]. More recently, they have been used to control the photochemical synthesis of Au and Ag nanoparticles in femtosecond-laser plasmas [21, 26, 73, 75]. The addition of $N_2O$ to aqueous $[AuCl_4]^-$ is expected to limit the availability of hydrated electrons and slow the $[AuCl_4]^-$ reduction rate, forming fewer Au(0) nuclei in solution. This situation would result in a significant number of Au(III) ions being reduced on the surface of the Au(0) nuclei in the presence of $H_2O_2$, promoting the surface-mediated autocatalytic growth into larger AuNPs. In contrast, the addition of a hydroxyl radical scavenger such as 2-propanol should not only limit the production of $H_2O_2$ via Eq. (8), but also prevent the quenching of hydrated electrons via Eq. (6). As a result, $[AuCl_4]^−$ reduction should be fast and autocatalytic AuNP growth should be limited, resulting in smaller AuNPs. These predictions have been laid out in recent literature [21, 26].

Belmouaddine et al. [26] investigated the effect of adding $N_2O$ or 2-propanol to aqueous $[AuCl_4]^−$ solutions they irradiated with 1.35 mJ, 112 fs pulses. They monitored the reduction kinetics to determine the $k_1$ and $k_2$ rate constants in Eq. (13). By comparing the $k_2/k_1$ ratios obtained in the two scavenger experiments, they were able to relate each scavenger to its role in the reduction and autocatalytic growth processes. In the presence of the hydrated-electron scavenger $N_2O$, the $k_2/k_1$ ratio was two orders of magnitude higher than it was when the radical scavenger 2-propanol was used. This is consistent with the dependence of $k_1$ and $k_2$ on hydrated-electrons and $H_2O_2$, discussed in Section 3. The resulting AuNPs synthesized in the presence of $N_2O$ and 2-propanol were $54.4 \pm 9.8$ nm, and $28.5 \pm 5.9$ nm. These results are consistent with the predictions that (1) slow nucleation and significant autocatalytic growth in the presence of $N_2O$ will produce large AuNPs, and (2) fast nucleation and limited autocatalytic growth in the presence of 2-propanol will produce small AuNPs.

In another study, Uwada et al. [21] investigated the effects of alcohols (1-propanol, 2-propanol, ethanol) on aqueous $[AuCl_4]^−$ solutions irradiated with 120 fs pulses, using a series of pulse energies from 1 to 50 $\mu$J. At low pulse energies, with intensity below $7 \times 10^{15}$ W cm$^{-2}$, no AuNPs formed if there were no alcohols. AuNP size-dependence on the pulse energy followed the *opposite* trend to that observed in Refs. [16, 20] and discussed in Section 4.1.2: the AuNPs formed in solutions containing 1-propanol increased in diameter from 24 to 37 nm when the intensity increased from $2 \times 10^{15}$ to $7 \times 10^{15}$ W cm$^{-2}$ [21]. The authors proposed that the alcohol radicals formed from the OH. scavenging reaction

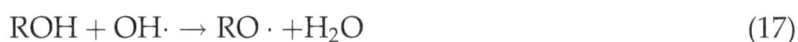

$$ROH + OH \cdot \rightarrow RO \cdot + H_2O \tag{17}$$

act as the primary reducing agents of $[AuCl_4]^−$ at low laser intensities where few hydrated electrons are formed. These results suggest that radical scavengers not only manage the AuNP size, but also boost photochemical reduction of $[AuCl_4]^−$ into AuNPs at lower laser intensities by providing an additional free-radical reducing agent.

### 4.2.2. pH

Changing the pH of irradiated aqueous $[AuCl_4]^−$ solutions by adding either HCl or KOH affects both the reduction kinetics and the resulting AuNP sizes [15]. Solution pH is well known to affect Au(III) complex speciation: $[AuCl_4]^−$ dominates under acidic conditions, $[Au(OH)_4]^−$ dominates under basic conditions, and mixtures of $[AuCl_x(OH)_{4-x}]^−$, $x = 1 - 3$, species exist under neutral conditions [76, 77]. Different complex stabilities were thought to be the driving force for Au(III) reduction with chemical reducing agents, where $[Au(OH)_4]^−$ was less reactive because of stronger Au-OH bonds, compared to Au-Cl bonds [76, 77]. With increasing pH, as solution was irradiated with 36 ps, 2.5 mJ pulses under SSTF focusing conditions, the reverse trend occurred, and higher $[AuCl_4]^-$ reduction rates formed smaller AuNPs [15]. At low pH, the hydrated-electron lifetime is reduced [55] and $H_2O_2$ oxidizes AuNPs [78], causing a slow $[AuCl_4]^-$ reduction rate that produced large, polydisperse $19.4 \pm 7.1$ nm AuNPs (at pH 2.5). When pH was higher (pH 5.4), the hydrated-electron lifetime is longer [55] and the oxidation potential of $H_2O_2$ increases as it is deprotonated to $HO_2^-$ [59], leading to faster reduction of $[AuCl_4]^-$ and small AuNPs with size distributions of $4.8 \pm 1.9$ nm. Slightly larger $6.6 \pm 3.1$ nm AuNPs were formed at pH 8.4 due to the acceleration of the autocatalytic growth rate constant $k_2$ in Eq. (13).

For comparison with the results in Ref. [15], experiments performed in our laboratory using the tight-focusing conditions in Ref. [16] also showed that the AuNP size depends on solution pH. Aqueous solutions (0.1 mM KAuCl$_4$) with varying amounts of KOH (up to 0.75 mM, pH 4.0–9.3) were irradiated with 50 $\mu$J, 30 fs pulses for 10–33 min, sufficient to convert all $[AuCl_4]^-$ to AuNPs. The UV–vis spectra recorded when the conversions of $[AuCl_4]^-$ to AuNPs were completed are shown in **Figure 7(a)**. As the solution pH increases, the SPR feature blue-shifts and decreases in intensity, indicating the production of smaller AuNPs [70]. TEM analysis of the AuNPs synthesized at pH 4.0, 5.2, and 9.3 (**Figure 7(b)–(d)**) agreed. At pH 4.0, there is a distinct bimodal size-distribution, with a number of small ($< 4$ nm) AuNPs and a broad distribution of particles to as large as 65 nm. As a result, a meaningless statistical size-distribution of $8.6 \pm 12.4$ nm is obtained. At pH 5.2, there is a bimodal distribution centered at 3 and 18 nm, with a size-distribution of $8.6 \pm 6.7$ nm. The most monodisperse AuNPs are seen at pH 9.3, with a size-distribution of $9.2 \pm 4.5$ nm. The slightly larger average AuNP size is because there are very few $< 4$ nm particles compared to what had been seen at lower pH. This absence of extremely small particles is likely due to the high concentration of $H_2O_2$ produced in the plasma, as discussed in Ref. [15].

### 4.2.3. Capping agents

One of the most widely used strategies for controlling AuNP size during chemical synthesis is the addition of capping agents, including polymers and surfactants, to stop particle growth [2]. Increasing the molar ratio of a capping agent to Au(III) salt typically results in smaller AuNPs in chemical syntheses [79]. A similar trend is observed for femtosecond laser-based syntheses of AuNPs, as shown in **Table 2**. Nakamura et al. [14] found that addition of polyvinylpyr-rolidone (PVP) to aqueous HAuCl$_4$ decreased AuNP size and gave a tighter size-distribution,

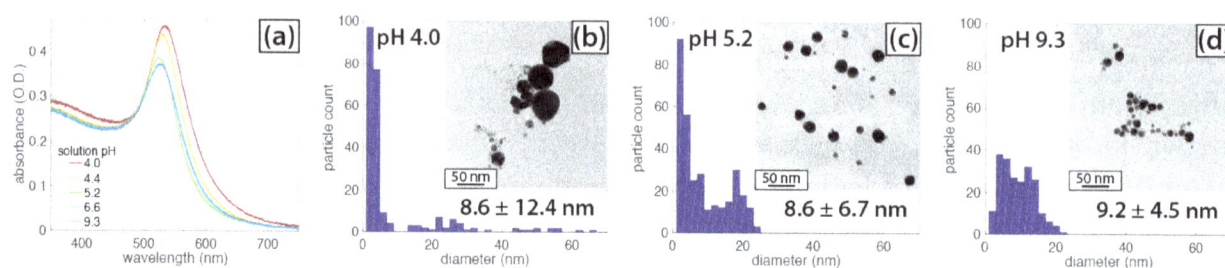

**Figure 7.** UV-vis spectra (a) and TEM images with size distributions (b)–(d) of AuNPs synthesized at different solution pH using the experimental setup in ref. [16].

| Ref. | Laser condition | Capping agent | Capping agent: Au | Size (nm) | Size range (nm) |
|---|---|---|---|---|---|
| [14] | High-NA | PVP[a] | 0.001 | 2, 7 | 1 − 77 |
| | | | 0.01 | 4.5 | 1 − 17 |
| | | | 0.1 | 3 | 2 − 7 |
| [19] | SSTF | PEG$_{45}$ | 0.05 | 11 ± 2.4 | 5 − 18 |
| | | | 0.1 | 6.0 ± 1.4 | 3 − 11 |
| | | | 0.25 | 3.9 ± 0.7 | 2 − 7 |
| | | | 0.5 | 4.1 ± 0.8 | 2 − 7 |
| [45] | Ablation, | Dextran[b] | 0.0002 | 76 ± 19 | |
| | Low-NA | | 0.002 | 30 ± 6 | |
| | | | 0.007 | 8.5 ± 1.7 | |
| | | | 0.05 | 3.0 ± 0.8 | |
| | | | 0.2 | 3.7 ± 1.1 | |

[a]Median size estimated from reported histogram.
[b]Standard deviation values estimated from 20 to 30% reported.

**Table 2.** Capping agents effect on AuNP size.

up to a 0.1:1 PVP:Au ratio. Tangeysh et al. [19] used poly(ethylene glycol) (PEG$_{45}$, $n = 45$) and found an optimal PEG$_{45}$:Au ratio of 0.25:1. The same trend is observed when femtosecond pulses are used to ablate an Au target [45], where the smallest particles made from irradiating AuNPs in solutions of varied dextran concentrations were optimized at a dextran:Au ratio of 0.05:1.

In addition to controlling AuNP size in femtosecond laser-based syntheses, capping agents like PEG$_{45}$ [19], chitosan (a cationic polysaccharide) [29], and (2-hydroxyethyl) trimethylammonium glycinate ([HETMA][Gly], an ionic liquid) [30] accelerate the conversion of [AuCl$_4$]$^-$ to AuNPs. It was proposed that fragmentation of PEG$_{45}$ in the laser-generated plasma produces alcohol radicals (analogous to Eq. (17)) that can reduce [AuCl$_4$]$^-$ and accelerate AuNP formation [19]. The formation of AuNPs in the presence of chitosan correlates with the oxidation of hydroxyl groups on the chitosan [29], indicating that the chitosan contributes to [AuCl$_4$]$^-$ reduction. The [HETMA][Gly] forms a complex with [AuCl$_4$]$^-$, which facilitated its reduction under OB conditions with tight-focusing [30]. These results suggest the potential of selective control over both [AuCl$_4$]$^-$ reduction kinetics and AuNP size through careful choice of capping agents and their concentrations.

## 5. Conclusion

Photochemical reduction of [AuCl$_4$]$^-$ using femtosecond laser irradiation is a simple, green method for controlling the growth of AuNPs. This chapter presented a review of the physical and chemical mechanisms of aqueous [AuCl$_4$]$^-$ transformation to AuNPs, from the physical processes occurring in plasma to AuNP size-control through selective tailoring of the solution composition. In Section 2, we discussed the physical processes of OB and SCE that occur because of ultrafast laser irradiation of water. The time-dependent electron density generated in OB plasma was modeled in relation to laser intensity, and the role that electrons play in $[AuCl_4]^-$ reduction were explained in Section 3. Reactive species produced in OB plasma were identified, and their roles in the kinetically controlled photochemical reduction of $[AuCl_4]^-$ (electrons $\propto k_1$) and surface-mediated autocatalytic growth into AuNPs ($H_2O_2$ production $\propto k_2$) were quantified and discussed. Finally, in Section 4, these approaches to control the size of AuNPs were reviewed. Both laser parameters (focusing-geometry, pulse energy, and duration) and solution modifications (pH, adding scavengers or capping agents) were discussed in how they affected the chemical system and reaction mechanism, allowing for size-control of AuNPs. Laser parameters and solution composition both play significant roles in the formation and resulting size of AuNPs, and this chapter highlights these considerations to direct future research.

## Acknowledgements

This work was supported by the American Chemical Society Petroleum Research Fund under Grant no 57799-DNI10 and Virginia Commonwealth University.

## Author details

Mallory G. John, Victoria Kathryn Meader and Katharine Moore Tibbetts*

*Address all correspondence to: kmtibbetts@vcu.edu

Department of Chemistry, Virginia Commonwealth University, Richmond, VA, USA

## References

[1]  Zhang JZ, Noguez C. Plasmonic optical properties and applications of metal nanostructures. Plasmonics. 2008;**3**:127

[2]  Dreaden EC, Alkilany AM, Huang X, Murphy CJ, El-Sayed MA. The golden age: Gold nanoparticles for biomedicine. Chemical Society Reviews. 2012;**41**:2740

[3]  Dreaden EC, Mackey MA, Huang X, Kang B, El-Sayed MA. Beating cancer in multiple ways using nanogold. Chemical Society Reviews. 2011;**40**:3391

[4]  Bastús NG, Comenge J, Puntes V. Kinetically controlled seeded growth synthesis of citrate-stabilized gold nanoparticles of up to 200 nm: Size focusing versus ostwald ripening. Langmuir. 2011;**27**:11098

[5]  Haynes CL, McFarland AD, Duyne RPV. Surface-Enhanced Raman Spectroscopy. Ana-

lytical Chemistry. 2005;**77**:338 A

[6] Alkilany AM, Lohse SE, Murphy CJ. The gold standard: Gold nanoparticle libraries to understand the nano-bio interface. Accounts of Chemical Research. 2013;**46**:650

[7] Mukherjee S, Libisch F, Large N, Neumann O, Brown LV, Cheng J, Lassiter JB, Carter EA, Nordlander P, Halas NJ. Hot electrons do the impossible: Plasmon-induced dissociation of $H_2$ on Au. Nano Letters. 2013;**13**:240

[8] Linic S, Christopher P, Ingram DB. Plasmonic-metal nanostructures for efficient conversion of solar to chemical energy. Nature Materials. 2011;**10**:911 dEP

[9] Xia Y, Xiong Y, Lim B, Skrabalak S. Shape-controlled synthesis of metal nanocrystals: Simple chemistry meets complex physics?. Angewandte Chemie, International Edition. 2009;**48**:60

[10] Personick ML, Mirkin CA. Making sense of the mayhem behind shape control in the synthesis of gold nanoparticles. Journal of the American Chemical Society. 2013;**135**:18238

[11] Zeng H, Du X-W, Singh SC, Kulinich SA, Yang S, He J, Cai W. Nanomaterials via laser ablation/irradiation in liquid: A review. Advanced Functional Materials. 2012;**22**:1333

[12] Svetlichnyi VA, Shabalina AV, Lapin IN, Goncharova DA. In: Yang D, editor. Metal oxide nanoparticle preparation by pulsed laser ablation of metallic targets in liquid. Applications of Laser Ablation - Thin Film Deposition, Nanomaterial Synthesis and Surface Modification. Rijeka: InTech; 2016 Chap. 11

[13] Sylvestre J-P, Poulin S, Kabashin AV, Sacher E, Meunier M, Luong JHT. Surface chemistry of gold nanoparticles produced by laser ablation in aqueous media. The Journal of Physical Chemistry B. 2004;**108**:16864

[14] Nakamura T, Mochidzuki Y, Sato S. Fabrication of gold nanoparticles in intense optical field by femtosecond laser irradiation of aqueous solution. Journal of Materials Research. 2008;**23**:968

[15] Tibbetts KM, Tangeysh B, Odhner JH, Levis RJ. Elucidating strong field photochemical reduction mechanisms of aqueous $[AuCl_4]^-$: Kinetics of multiphoton photolysis and radical-mediated reduction. The Journal of Physical Chemistry. A. 2016;**120**:3562

[16] Meader VK, John MG, Rodrigues CJ, Tibbetts KM. Roles of free electrons and $H_2O_2$ in the optical breakdown-induced photochemical reduction of aqueous $[AuCl_4]^-$. The Journal of Physical Chemistry. A. 2017;**121**:6742

[17] Zhao C, Qu S, Qiu J, Zhu C. Photoinduced formation of colloidal Au by a near-infrared femtosecond laser. Journal of Materials Research. 2003;**18**:1710

[18] Nakamura T, Herbani Y, Ursescu D, Banici R, Dabu RV, Sato S. Spectroscopic study of gold nanoparticle formation through high intensity laser irradiation of solution. AIP Advances. 2013;**3**:082101

[19] Tangeysh B, Moore Tibbetts K, Odhner JH, Wayland BB, Levis RJ. Gold nanoparticle

synthesis using spatially and temporally shaped femtosecond laser pulses: Post-irradiation auto-reduction of aqueous $[AuCl_4]^-$. Journal of Physical Chemistry C. 2013;**117**:18719

[20] Odhner JH, Moore Tibbetts K, Tangeysh B, Wayland BB, Levis RJ. Mechanism of improved Au nanoparticle size distributions using simultaneous spatial and temporal focusing for femtosecond laser irradiation of aqueous $KAuCl_4$. Journal of Physical Chemistry C. 2014; **118**:23986

[21] Uwada T, Wang S-F, Liu T-H, Masuhara H. Preparation and micropatterning of gold nanoparticles by femtosecond laser-induced optical breakdown. Journal of Photochemistry and Photobiology A. 2017;**346**:177

[22] Herbani Y, Nakamura T, Sato S. Synthesis of platinum-based binary and ternary alloy nanoparticles in an intense laser field. Journal of Colloid and Interface Science. 2012;**375**:78

[23] Sarker MSI, Nakamura T, Herbani Y, Sato S. Fabrication of Rh based solid-solution bimetallic alloy nanoparticles with fully-tunable composition through femtosecond laser irradiation in aqueous solution. Applied Physics A: Materials Science & Processing. 2012; **110**:145

[24] Tangeysh B, Moore Tibbetts K, Odhner JH, Wayland BB, Levis RJ. Triangular gold nanoplate growth by oriented attachment of Au seeds generated by strong field laser reduction. Nano Letters. 2015;**15**:3377

[25] Tangeysh B, Tibbetts KM, Odhner JH, Wayland BB, Levis RJ. Gold nanotriangle formation through strong-field laser processing of aqueous $KAuCl_4$ and postirradiation reduction by hydrogen peroxide. Langmuir. 2017;**33**:243

[26] Belmouaddine H, Shi M, Karsenti P-L, Meesat R, Sanche L, Houde D. Dense ionization and subsequent non-homogeneous radical-mediated chemistry of femtosecond laser-induced low density plasma in aqueous solutions: Synthesis of colloidal gold. Physical Chemistry Chemical Physics. 2017;**19**:7897

[27] Nakashima N, Yamanaka K, Saeki M, Ohba H, Taniguchi S, Yatsuhashi T. Metal ion reductions by femtosecond laser pulses with micro-joule energy and their efficiencies. Journal of Photochemistry and Photobiology A. 2016;**319–320**:70

[28] Rodrigues CJ, Bobb J, John MG, Fisenko S, El-Shal MSI, Tibbetts KM. Photochemical reduction of $[AuCl_4]^-$ with nanosecond and femtosecond laser pulses: Electrons or heat?. 2018. in preparation

[29] Ferreira PHD, Vivas MG, Boni LD, dos Santos DS, Balogh DT, Misoguti L, Mendonca CR. Femtosecond laser induced synthesis of Au nanoparticles mediated by chitosan. Optics Express. 2012;**20**:518

[30] Lu W-E, Zheng M-L, Chen W-Q, Zhao Z-S, Duan X-M. Gold nanoparticles prepared by glycinate ionic liquid assisted multi-photon photoreduction. Physical Chemistry Chemical Physics. 2012;**14**:11930

[31] Keldysh L. Ionization in field of a strong electromagnetic wave. Soviet Physics - JETP. 1965;**20**:1307

[32]  Shen YR. The Principles of Nonlinear Optics. New York: Wiley; 1984

[33]  Noack J, Vogel A. Laser-induced plasma formation in water at nanosecond to femtosecond time scales: Calculation of thresholds, absorption coefficients, and energy density. IEEE Journal of Quantum Electronics. 1999;**35**:1156

[34]  Vogel A, Noack J, Hüttman G, Paltauf G. Mechanisms of femtosecond laser nanosurgery of cells and tissues. Applied Physics B. 2005;**81**:1015

[35]  Liu W, Kosareva O, Golubtsov I, Iwasaki A, Becker A, Kandidov V, Chin S. Femtosecond laser pulse filamentation versus optical breakdown in $H_2O$. Applied Physics B: Lasers and Optics. 2003;**76**:215

[36]  Kandidov V, Kosareva O, Golubtsov I, Liu W, Becker A, Akozbek N, Bowden C, Chin S. Self-transformation of a powerful femtosecond laser pulse into a white-light laser pulse in bulk optical media (or supercontinuum generation). Applied Physics B: Lasers and Optics. 2003;**77**:149

[37]  Couairon A, Mysyrowicz A. Femtosecond filamentation in transparent media. Physics Reports. 2007;**441**:47

[38]  Linz N, Freidank S, Liang X-X, Vogel A. Wavelength dependence of femtosecond laser-induced breakdown in water and implications for laser surgery. Physical Review B. 2016;**94**:024113

[39]  Elles CG, Jailaubekov AE, Crowell RA, Bradforth SE. Excitation-energy dependence of the mechanism for two-photon ionization of liquid $H_2O$ and $D_2O$ from 8.3 to 12.4 eV. The Journal of Chemical Physics. 2006;**125**:044515

[40]  Linz N, Freidank S, Liang X-X, Vogelmann H, Trickl T, Vogel A. Wavelength dependence of nanosecond infrared laser-induced breakdown in water: Evidence for multiphoton initiation via an intermediate state. Physical Review B. 2015;**91**:134134

[41]  Fan CH, Sun J, Longtin JP. Breakdown threshold and localized electron density in water induced by ultrashort laser pulses. Journal of Applied Physics. 2002;**91**:2530

[42]  Efimenko ES, Malkov YA, Murzanev AA, Stepanov AN. Femtosecond laser pulse-induced breakdown of a single water microdroplet. Journal of the Optical Society of America B: Optical Physics. 2014;**31**:534

[43]  Marburger J. Self-focusing: Theory. Progress in Quantum Electronics. 1975;**4**:35

[44]  Brodeur A, Chin SL. Band-gap dependence of the ultrafast white-light continuum. Physical Review Letters. 1998;**80**:4406

[45]  Besner S, Kabashin AV, Winnik FM, Meunier M. Synthesis of size-tunable polymer-protected gold nanoparticles by femtosecond laser-based ablation and seed growth. Journal of Physical Chemistry C. 2009;**113**:9526

[46]  Besner S, Kabashin AV, Meunier M. Fragmentation of colloidal nanoparticles by femtosecond laser-induced supercontinuum generation. Applied Physics Letters. 2006;**89**:233122

[47]  Besner S, Kabashin A, Meunier M. Two-step femtosecond laser ablation-based method for the synthesis of stable and ultra-pure gold nanoparticles in water. Applied Physics A: Materials Science & Processing. 2007;**88**:269

[48]  Minardi S, Gopal A, Tatarakis M, Couairon A, Tamošauskas G, Piskarskas R, Dubietis A, Trapani PD. Time-resolved refractive index and absorption mapping of light-plasma filaments in water. Optics Letters. 2008;**33**:86

[49]  Minardi S, Gopal A, Couairon A, Tamoašuskas G, Piskarskas R, Dubietis A, Trapani PD. Accurate retrieval of pulse-splitting dynamics of a femtosecond filament in water by time-resolved shadowgraphy. Optics Letters. 2009;**34**:3020

[50]  Besner S, Meunier M. Femtosecond laser synthesis of AuAg nanoalloys: Photoinduced oxidation and ions release. Journal of Physical Chemistry C. 2010;**114**:10403

[51]  Chin SL, Lagacé S. Generation of $H_2$, $O_2$, and $H_2O_2$ from water by the use of intense femtosecond laser pulses and the possibility of laser sterilization. Applied Optics. 1996;**35**:907

[52]  Kurihara K, Kizling J, Stenius P, Fendler JH. Laser and pulse radiolytically induced colloidal gold formation in water and in water-in-oil microemulsions. Journal of the American Chemical Society. 1983;**105**:2574

[53]  Crowell RA, Bartels DM. Multiphoton ionization of liquid water with 3.0–5.0 eV photon. The Journal of Physical Chemistry. 1996;**100**:17940

[54]  Reuther A, Laubereau A, Nikogosyan DN. Primary photochemical processes in water. The Journal of Physical Chemistry. 1996;**100**:16794

[55]  Pommeret S, Gobert F, Mostafavi M, Lampre I, Mialocq J-C. Femtochemistry of the hydrated electron at decimolar concentration. The Journal of Physical Chemistry. A. 2001;**105**:11400

[56]  Nikogosyan DN, Oraevsky AA, Rupasov VI. Two-photon ionization and dissociation of liquid water by powerful laser UV radiation. Chemical Physics. 1983;**77**:131

[57]  Behar D, Rabani J. Kinetics of hydrogen production upon reduction of aqueous $TiO_2$ nanoparticles catalyzed by $Pd^0$, $Pt^0$, or $Au^0$ coatings and an unusual hydrogen abstraction; steady state and pulse radiolysis study. The Journal of Physical Chemistry. B. 2006;**110**:8750

[58]  Zayats M, Baron R, Popov I, Willner I. Biocatalytic growth of Au nanoparticles: From mechanistic aspects to biosensors design. Nano Letters. 2005;**5**:21

[59]  McGilvray KL, Granger J, Correia M, Banks JT, Scaiano JC. Opportunistic use of tetrachloroaurate photolysis in the generation of reductive species for the production of gold nanostructures. Physical Chemistry Chemical Physics. 2011;**13**:11914

[60]  LaMer VK, Dinegar RH. Theory, production and mechanism of formation of monodispersed hydrosols. Journal of the American Chemical Society. 1950;**72**:4847

[61]  Yan H, Cingarapu S, Klabunde KJ, Chakrabarti A, Sorensen CM. Nucleation of gold

nanoparticle superclusters from solution. Physical Review Letters. 2009;**102**:095501

[62] Turkevich J, Stevenson PC, Hillier J. A Study of the nucleation and growth processes in the synthesis of colloidal gold. Discussions of the Faraday Society. 1951;**11**:55

[63] Takiyama K. Formation and aging of precipitates. VIII. Formation of monodisperse particles (I) Gold sol particles by sodium citrate method. Bulletin of the Chemical Society of Japan. 1958;**31**:944

[64] Esumi K, Hosoya T, Suzuki A, Torigoe K. Spontaneous formation of gold nanoparticles in aqueous solution of sugar-persubstituted poly(amidoamine)dendrimers. Langmuir. 2000;**16**:2978

[65] Shields SP, Richards VN, Buhro WE. Nucleation control of size and dispersity in aggregative nanoparticle growth. A study of the coarsening kinetics of thiolate-capped gold nanocrystals. Chemistry of Materials. 2010;**22**:3212

[66] Njoki PN, Luo J, Kamundi MM, Lim S, Zhong C-J. Aggregative growth in the size-controlled growth of monodispersed gold nanoparticles. Langmuir. 2010;**26**:13622

[67] Watzky MA, Finke RG. Transition metal nanocluster formation kinetic and mechanistic studies. A new mechanism when hydrogen is the reductant: Slow, continuous nucleation and fast autocatalytic surface growth. Journal of the American Chemical Society. 1997;**119**:10382

[68] Finney EE, Finke RG. Nanocluster nucleation and growth kinetic and mechanistic studies: A review emphasizing transition-metal nanoclusters. Journal of Colloid and Interface Science. 2008;**317**:351

[69] Piella J, Bastús NG, Puntes V. Size-Controlled Synthesis of Sub-10-nanometer Citrate-Stabilized Gold Nanoparticles and Related Optical Properties. Chemistry of Materials. 2016;**28**:1066

[70] Haiss W, Thanh NTK, Aveyard J, Fernig DG. Determination of size and concentration of gold nanoparticles from UV-vis spectra. Analytical Chemistry. 2007;**79**:4215

[71] Oron D, Silberberg Y. Spatiotemporal coherent control using shaped, temporally focused pulses. Optics Express. 2005;**13**:9903

[72] Faccio D, Tamošauskas G, Rubino E, Darginavičius J, Papazoglou DG, Tzortzakis S, Couairon A, Dubietis A. Cavitation dynamics and directional microbubble ejection induced by intense femtosecond laser pulses in liquids. Physical Review E. 2012;**86**:036304

[73] Meader VK, John MG, Frias Batista LM, Ahsan S, Tibbetts KM. Radical chemistry in a femtosecond laser plasma: Photochemical reduction of $Ag^+$ in liquid ammonia solution. Molecules. 2018;**23**:532

[74] LaVerne JA, Pimblott SM. Scavenger and time dependences of radicals and molecular products in the electron radiolysis of water: Examination of experiments and models. The Journal of Physical Chemistry. 1991;**95**:3196

[75] Herbani Y, Nakamura T, Sato S. Silver nanoparticle formation by femtosecond laser

induced reduction of ammonia-containing AgNO$_3$ solution. Journal of Physics Conference Series. 2017;**817**:012048

[76] Goia D, Matijevi'c E. Tailoring the particle size of monodispersed colloidal gold. Colloids and Surfaces, A: Physicochemical and Engineering Aspects. 1999;**146**:139

[77] Wang S, Qian K, Bi X, Huang W. Influence of speciation of aqueous HAuCl$_4$ on the synthesis, structure, and property of Au colloids. Journal of Physical Chemistry C. 2009; **113**:6505

[78] Ni W, Kou X, Yang Z, Wang J. Tailoring longitudinal surface plasmon wavelengths, scattering and absorption cross sections of gold nanorods. ACS Nano. 2008;**2**:677

[79] Oh E, Susumu K, Goswami R, Mattoussi H. One-phase synthesis of water-soluble gold nanoparticles with control over size and surface functionalities. Langmuir. 2010;**26**:7604

# 3

# Photochemical Degradation Processes of Painting Materials from Cultural Heritage

Rodica-Mariana Ion, Alexandrina Nuta,
Ana-Alexandra Sorescu and Lorena Iancu

## Abstract

This chapter describes some recent studies and applications of photochemistry in the physical–chemical characterization of two acrylic paint materials based on phthalocyanines and the study of the photodegradation (photobleaching) processes which could occur, caused by exposure to artificial irradiation, similar as in the museum. The studies in this paper has been conducted on phthalocyanines, these compounds being known as organic colorants in painting. Their color depends not only on the chemical nature of the colorant, which play an important role in the kinetics and degree of aging, but also on the compounds added to the paints ($TiO_2$, micas, arylamide yellow). The techniques used in such studies involve UV–Vis spectroscopy, gloss, and colorimetric measurements, comparing our results with similar ones from the literature.

**Keywords:** phthalocyanine, absorption, photobleaching, artifacts

## 1. Introduction

Photochemistry is the science that deals with the study of physical and chemical processes that arise from the interaction between radiation used for irradiation and absorbing molecules [1].

The light radiation together in the presence of oxygen induces photooxidative reactions on synthetic organic materials, mostly based on chain scission and cross-linking reactions of polymers used as binders in artists' paints, such as acrylics [2]. Photodegradation reactions cause changes in the physical and mechanical properties of the materials such as yellowing phenomena, cracking, embrittlement, and stiffening of the paint films as well as changes in solubility. The general photodegradation processes on polymers are initiated by the absorption

| Pigment CI name | Pigment chemical name | Paint marketing name | Structure |
|---|---|---|---|
| **PB15:1** | *Alpha copper phthalocyanine* (1935) | Winsor Blue RS<br>Phthalo Blue RS<br>Phthalo Blue Red<br>Phthalo Blue<br>Berlin Blue | |
| **PB15:3** | *Beta copper phthalocyanine* (1933, 1935) | Phthalocyanine Blue<br>Phthalo Blue<br>Blockx Blue<br>Winsor Blue GS<br>Phthalo Blue Green<br>Phthalo Blue GS<br>Primary Blue<br>Primary Blue-Cyan<br>Manganese Blue Hue | |
| **PB15:6** | *Epsilon copper phthalocyanine* (1935) | Phthalo Blue (red shade)<br>Helio Blue RS | |
| **PB16** | *Metal-free phthalocyanine* (1936) | Turquoise Green<br>Phthalo Turquoise<br>Marine Blue<br>Caribbean Blue | |

| Pigment CI name | Pigment chemical name | Paint marketing name | Structure |
|---|---|---|---|
| PB17 | *Trisulphonated copper phthalocyanine* (1935) | Peacock Blue [discontinued in 2005] | |
| PG7 | *Chlorinated copper phthalocyanine* (1927, 1938) | Winsor Green BS Phthalocyanine Green Phthalo Green BS Cupric Green Deep Phthalo Green Blockx green | |
| PG36 | *Chlorobrominated copper phthalocyanine* (1938) | Phthalo Green YS Cupric Green light Bright Green Bamboo Green Winsor Green YS Phthalocyanine Green (yellow shade) | |

**Table 1.** The structure of the Phthalocyanine blue and Phthalocyanine green pigments.

of photons by the molecules [3]. The excited molecule loses the absorbed energy through initiation, propagation, and termination reactions. In case of acrylates, two processes, chain scission and crosslinking, are competing, depending on the length of the side groups [4].

The carbonyl groups from the acrylic structure are also sensitive to secondary photodegradation reactions, such as Norrish reactions [5]. The photooxidation takes generally place at first in the uppermost layer of the paints and proceeds toward the bulk, depending on the radiation, oxygen diffusion, time of exposure, and the characteristics of the materials exposed to the radiation. The additives present in the paint formulations might interfere with the photo-oxidative reactions and could act by catalyzing and promoting or preventing photodegradation reactions, depending on their chemical properties and color [6].

In the present chapter, the influence of different types of phthalocyanines, such as Phthalo Blue and Phthalo Green (**Table 1**), on the photostability of acrylic paints [7] is studied. Investigations were carried out on unaged and artificially aged self-made and commercially available acrylic paints. Thus, paints were exposed to artificial accelerated aging using a Hg medium-pressure lamp, with spectral range comparable to outdoor solar radiation.

For the identification and characterization of acrylic films could be mentioned some non-invasive techniques, as: infrared spectroscopy Fourier-transformed (FTIR) in the attenuated total reflection (ATR) mode, UV–Vis spectroscopy, gloss and colorimetric measurements, all these techniques being useful for color changes and ageing processes investigations.

Taking into account that Phthalo Blue and Phthalo Green are the trade names for phthalocyanine pigments which are widely found in original modern and contemporary artworks [8], in the present work, phthalocyanines containing copper (CuPc), having Color Index (CI) name Pigment Blue 15 (PB15), and metal-free phthalocyanine ($H_2Pc$), PB16, are discussed. The PB15 can present polymorphism (PB15:x), and each crystalline form is characterized by different chemical and color properties [9]. The green phthalocyanines are copper-based phthalocyanines which are completely chlorinated (Pigment Green 7) or chlorinated and brominated (Pigment Green 36) (**Table 1**).

## 2. General aspects about phthalocyanines

In 1935, the Imperial Chemical Industries began to manufacture copper phthalocyanine, and such compounds were introduced in the commerce with the name of "Monastral Fast Blue" at a London fair in November 1935, and from 1936 produced also in the USA, under other names [10]. In 1936 IG Farbenindustrie began to produce copper phthalocyanine at Ludwigshafen, and in the late 1930s, the DuPont Company, at Deepwater Point, New Jersey, began to produce copper phthalocyanine; the Standard Ultramarine and Color Company started its production in 1949 [11]. Particularly, since the end of the Second World War, phthalocyanine pigments became widely used [12]. Commercially, available pigments of different producers (Maimeri, Winsor & Newton, Lefranc & Bourgeois, etc.) containing phthalocyanines PB15:1, PB15:3, and PG7 were chosen and applied on different kinds of supports, as cardboard. Samples of pigment were prepared using acrylic colors. Phthalocyanine molecules (Pc) are composed of four indole units—pyrrole rings linked by nitrogen atoms conjugated with benzene rings. They have an extensively conjugated aromatic chromophore, exhibit UV–Vis absorption spectra with intense $\pi$-$\pi^*$ transitions [13]. They have an increased aromatic character which explains the intense near-infrared absorption of these compounds. Metalation reduces the electron density at the inner nitrogen atoms and in UV–Vis spectra produces a hypsochromic shift which depends on the electronegativity of the metal.

The stability of metal phthalocyanines is due to formation of four equivalent N → metal bonds involving filling of vacant $ns$, $np$, and $(n-1)$d or $n$d orbitals of the cation with electrons of the central nitrogen atoms.

Dyes and pigments are used in the mixture with polymer materials to provide color-changing properties. Under such context, Cu-phthalocyanine dye can help stabilize against degradation but in other situations such as photochemical aging can actually accelerate degradation. The excited Cu-phthalocyanine may abstract hydrogen atoms from methyl groups in the Pc, which increase the formation of free radicals. This acts as the starting points for the sequential photooxidation reactions leading to the degradation of the phthalocyanine [14].

Electron transfer sensitization is a mechanism where the excited Cu-phthalocyanine abstracts electrons from Pc to form Cu-Pc radical anion and Pc radical cations. These species in the presence of oxygen can cause oxidation of the aromatic ring [15].

## 3. Light effects on pigments

### 3.1. Photophysical aspects

Light is the electromagnetic radiation released by an external source of energy. Photochemistry uses radiation that has an energy between 1 eV and 1 KeV, corresponding to visible radiation with $\lambda = 400–700$ nm and ultraviolet radiation with $\lambda = 200–400$ nm. Radiation used in photochemistry is selectively absorbed by the molecules with which it interacts.

To initiate a photochemical reaction, the formation of an electronically excited state in a molecule is decisive. Such an excited molecule can be regarded as a new species, characterized by properties other than those of the same basic molecule. The excited-state molecule has a certain electronic distribution responsible for the specific chemical reactivity compared to the fundamental chemical state. The essence of an organic photochemical reaction is the activation of the molecules produced by the absorption of a photon. The main processes that occur during the light action on paints are identified in Jablonski diagram (**Figure 1**):

(a) **Non-radiative processes**, where the excited-state species release the excess of energy as heat by three different processes: *vibrational relaxation (VR)*, through which the excited molecule decreases its vibrational energy within a single electronic state, *internal conversion (IC)*; meaning a transition between two electronic states with the same spin multiplicity, generally followed by vibrational relaxation; and *intersystem crossing (ISC)*, which involves the transition between two electronic states with different spin multiplicity, generally followed by vibrational relaxation.

(b) **Radiative processes.** The excited-state species release the excess of energy as electromagnetic radiation. There are two known processes: *fluorescence (F)*, as spontaneous emission of radiation upon transition between two electronic states with the same spin multiplicity, and *phosphorescence (P)*, known as a spontaneous emission of radiation upon transition between two electronic states with different spin multiplicity.

(c) **Other deactivation processes, identified by** photochemical or photophysical reactions which compete the excited-state molecules.

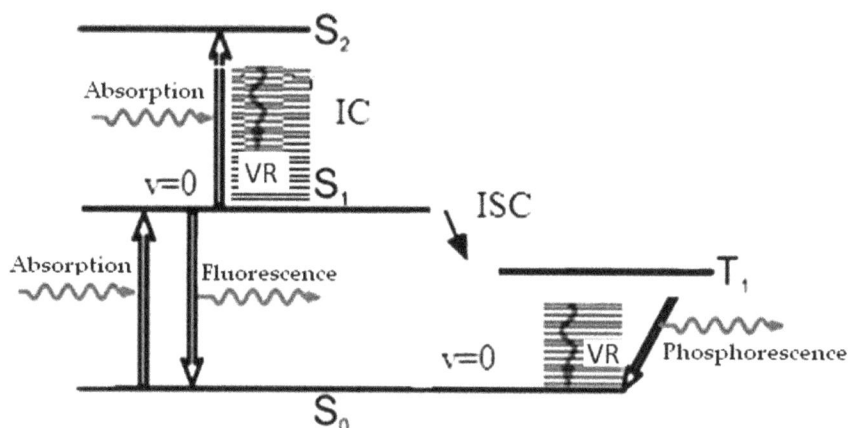

**Figure 1.** Jablonski diagram.

## 3.2. Photooxidation reaction

The photooxidation is a degradation of a compound in the presence of oxygen or ozone. The initiation of photooxidation reactions is due to the existence of chromophore groups in the molecules. For example, in a polymer case, by photooxidation, aldehydes, ketones, and carboxylic acids along or at the end of polymer chains are generated. The photooxidation reaction induces a chemical change that reduces the polymer's molecular weight. The material will become more brittle, with a reduction in its tensile, impact, and elongation strength. Also, a discoloration and loss of surface smoothness accompany photooxidation. High temperature significantly increases the effect of photooxidation.

The photooxidation reactions include chain scission, cross-linking, and secondary oxidative reactions, which obey the following process steps that can be considered [16], as follows (**Figure 2**):

1.  Initial step: Free radicals are formed by photon absorption.

2.  Chain propagation step: A free radical reacts with oxygen to produce a polymer peroxy radical (POO•). This reacts with a polymer molecule to generate polymer hydroperoxide (POOH) and a new polymer alkyl radical (P•).

3.  Chain branching: Polymer oxy radicals (PO•) and hydroxy radicals (HO•) are formed by photolysis.

4.  Termination step: Cross-linking is a result of the reaction of different free radicals with each other.

From mechanistic point of view, during the photooxidation two main reaction types occur (**Figure 3**):

**Type I mechanism:** a photosensitizer (PS) in its singlet or triplet excited state reacts with a substrate via (a) electron transfer or (b) hydrogen abstraction to yield free radicals, which will readily react with oxygen to form peroxide radicals and in turn start a radical chain reaction. These radicals react with oxygen to form reactive organic species (ROS).

**Type II mechanism:** in this process, the sensitizer in its excited state (commonly in its triplet state) transfers its energy to ground-state molecular oxygen, giving rise to the PS in its ground state and singlet oxygen ($^1O_2$), a very reactive oxygen species. On type II reaction, an energy transfer during a collision between the excited photosensitizer and molecular oxygen occurs, as it is reported in the literature's report. Usually, a photosensitizer, not degraded through photobleaching by $^1O_2$ or other processes, can produce $10^3$–$10^5$ molecules of singlet oxygen [19].

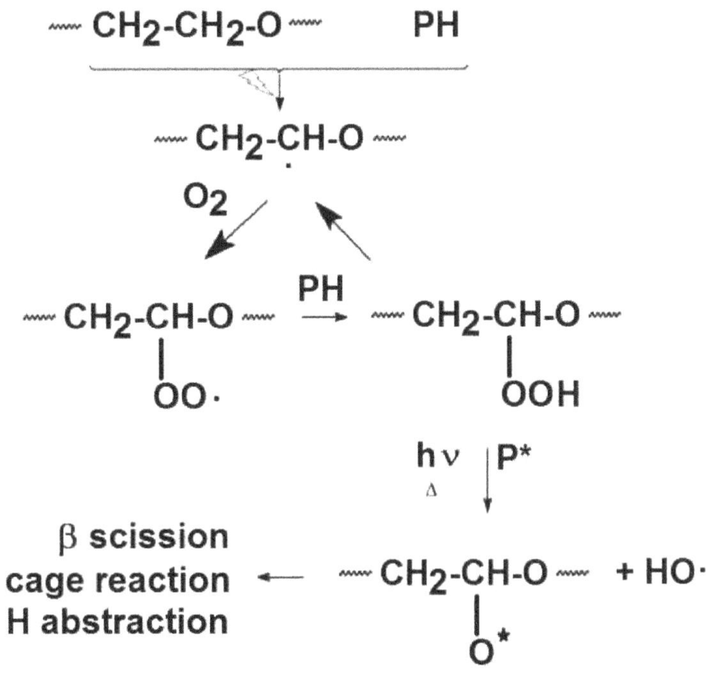

**Figure 2.** Adopted after [17, 18].

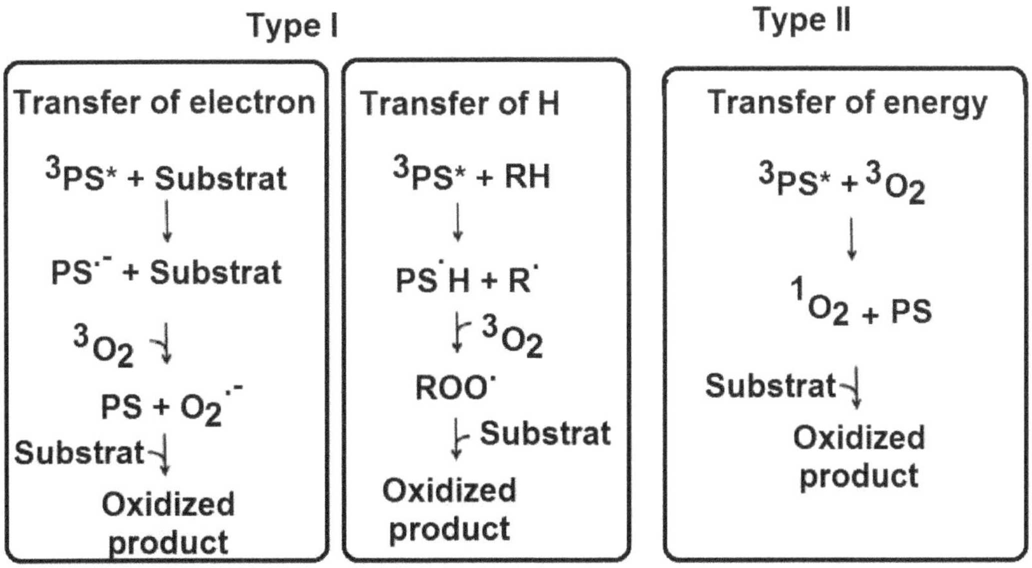

**Figure 3.** Mechanism of the photo-oxidation reaction.

# 4. Mechanism of phthalocyanine oxidative photobleaching

The main factors, which affect the photobleaching, are:

- Triplet excited state with long-lived triplet states and high triplet quantum yields induce a higher rate of photobleaching [20].

- Radiation type, the ultraviolet light being more effective in promoting phthalocyanine photobleaching than Q-band excitation [21].

The presence of oxygen favors the bleaching of metallic phthalocyanines (MPcs) predominantly due to photooxidation [22]. The oxygen concentration in oxygen-saturated solutions is $[O_2] = 10^{-3}–10^{-2}$ mol/l. The reaction between $^3Pc^*$ and oxygen is diffusion-controlled reaction, so that it is very rapid ($k_{O_2} = 10^9–10^{10}$ l/(mol. s)). A **type I photooxidation reaction** occurs only if $k_A[A]$ has a value of $10^6–10^7$ s$^{-1}$. If $k_A$ or $[A]$ has small values, then **type II reaction** occurs (if oxygen is present in small concentration). If $k_A$ has a great value, then **type III reaction** occurs (the oxygen is completely absent in the system).

## 4.1. Photooxidation study case

The tested compounds in this chapter were Phthalo Blues (CI names PB15:1, PB15:3, PB15:6, and PB16) and Phthalo Green (PG7) with the chemical structure shown in **Table 2**; all these structures have been exposed to irradiation with polychromatic light generated by a Hg medium lamp 375 nm.

| Acrylic paints | Market name | Pigment |
|---|---|---|
| 143 | Phthalo Blue | Phthalocyanine Blue PB15 |
| 159 | Primary Blue | Phthalocyanine Blue PB15 |
| 718 | Metallic Blue | Titanium dioxide-coated mica PW6; Phthalocyanine Blue PB15 |
| 335 | Emerald Green | Arylamide Yellow 10G PY3; Phthalocyanine Green PG7 |
| 375 | Sap Green | Arylamide Yellow 5GX PY 74; Phthalocyanine Blue PB15 |
| 719 | Metallic Green | Titanium dioxide-coated mica PW6; Phthalocyanine Green PG7 |
| 154 | Phthalo Turquoise | Phthalocyanine Blue PB15; Phthalocyanine Green PG7 |
| 343 | Hookers Green | Phthalocyanine Green PG7; Iron oxide PR101; Arylamide Yellow 5GX PY 74 |
| 386 | Phthalo Green | Phthalocyanine Green PG7 |

Table 2. The composition of the tested acrylic paints.

The tested samples are shown in **Figure 4**.

A low photostability of Phthalo Green compared with Phthalo Blue has been observed, similar with the photodegradation study of the Heliogen Blau 6900 and Heliogen Grün L8730 standard solutions (**Figures 5** and **6**).

For all tested samples, measurements of gloss degree and optical parameters ($\Delta Ex^*$ and $\Delta bx^*$) were performed. The variation of these parameters was also measured during the photodegradation process with polychromatic light irradiation provided by an artificial light source similar to solar radiation spectrum. After analyzing the obtained results, it was possible to assign that the gloss degree decreases gradually after irradiation with polychromatic radiation (at the samples 375, 335, 154, 343, and 386), while at the other samples (718, 719, 143, 159),

**Figure 4.** The cardboard painted with blue and green pigments.

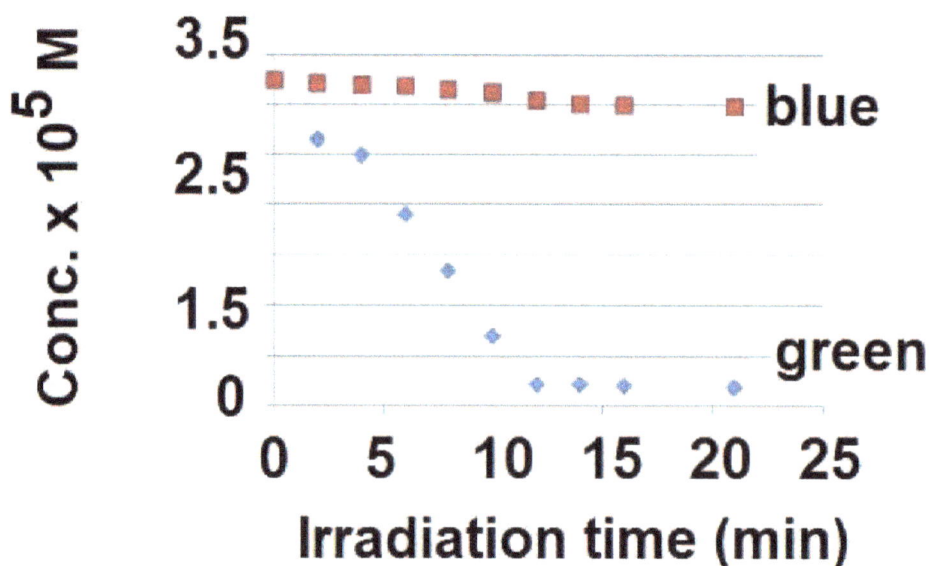

**Figure 5.** Photodegradation kinetics of Heliogen Blue 6900 and Heliogen Grün L8730 standard solutions.

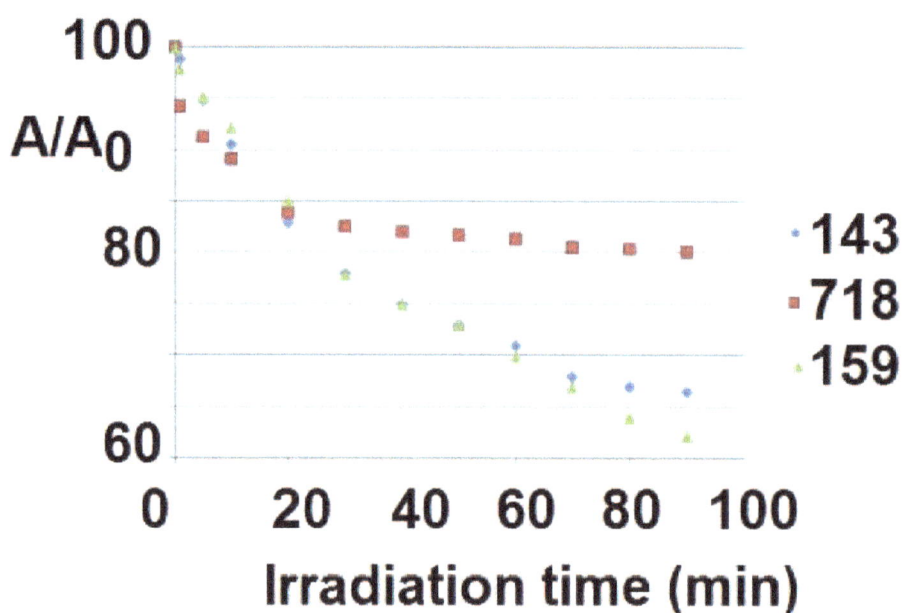

**Figure 6.** Photodegradation kinetics of some acrylic paints.

there is a significant decrease in this parameter recorded in the first 30 min, followed by a new slight increase of gloss degree. In the case of acrylic paints containing $TiO_2$ and micas, a decrease of the gloss parameter after the first 30 min of irradiation was observed, most probably due to the role of $TiO_2$, recognized as photocatalyst which generates OH radicals and, in this case a good contributor to photocatalytic degradation processes (**Figures 7** and **8**). In the case of the paints that contain arylamide yellow, a limited stability of these paints is observed, due to this pigment light sensitivity. This behavior is proven by a low increase gloss index for the paints that contain arylamide yellow (**Figure 7**).

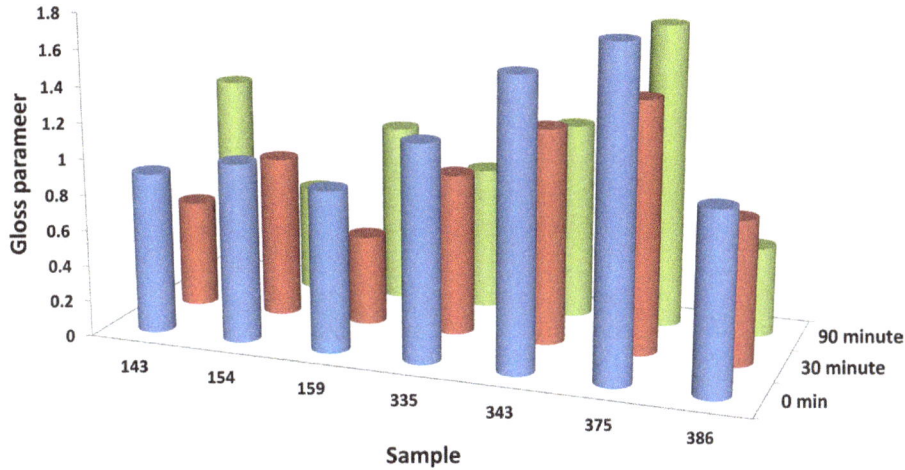

**Figure 7.** Gloss parameter variation.

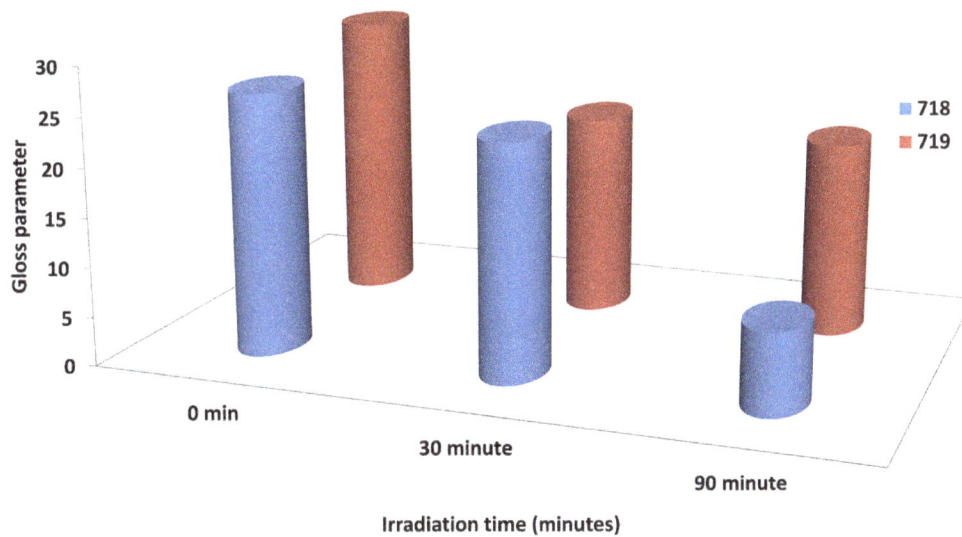

**Figure 8.** Gloss parameter variation in case of acrylic paints containing $TiO_2$ and micas.

The same trends are recorded for the $\Delta Ex^*$ and $\Delta bx^*$ optical measurements, **Figures 9** and **10** for $\Delta Ex^*$ and **Figures 11** and **12** for $\Delta bx^*$.

With regard to the change of the sample yellowness ($\Delta b^* = |b_f^* - b_i^*|$), the following classes are specified: stable for $\Delta b^* \leq 3$ points of absolute increase, moderately stable for $\Delta b^* > 3$ and $\Delta b^* \leq 8$ points of absolute increase, and unstable for $\Delta b^* > 8$ points of absolute increase. In this way, at the samples 143, 159, 718, and 719 (samples that contains $TiO_2$, micas, arylamide yellow), the yellowness corresponds to an increase of the unstability ($\Delta b^* > 8$), obeying the abovementioned rule, while for the samples 375, 335, 154, 343, and 386 ($\Delta b^* > 3$ and $\leq 8$), the yellowness corresponds to a moderate stability, as shown in **Figures 11** and **12**. For all the paints, subsequent light exposure resulted in a more severe degradation with large deterioration in optical properties. However, this tendency is the same as for $\Delta b^*$ parameter: moderately stable for all the samples, except the samples 143, 159, 718, and 719, with a high instability.

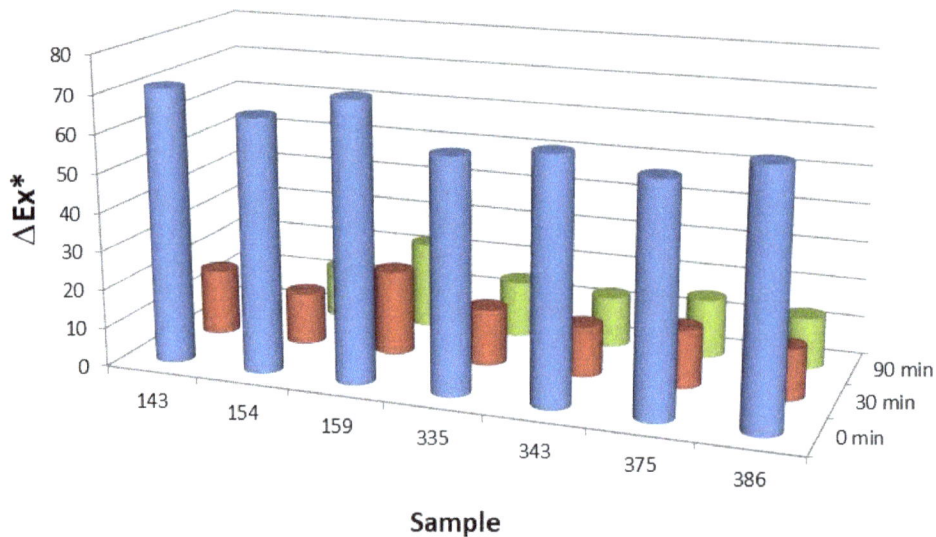

**Figure 9.** Changes of ΔEx* for irradiated samples at different times.

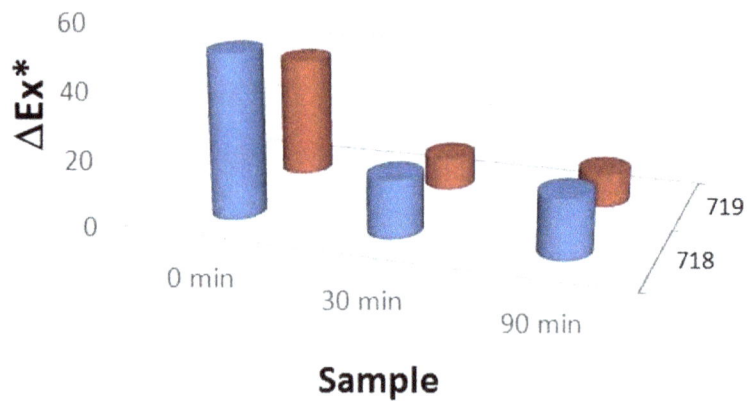

**Figure 10.** Changes of ΔEx* for irradiated metallic samples at different times.

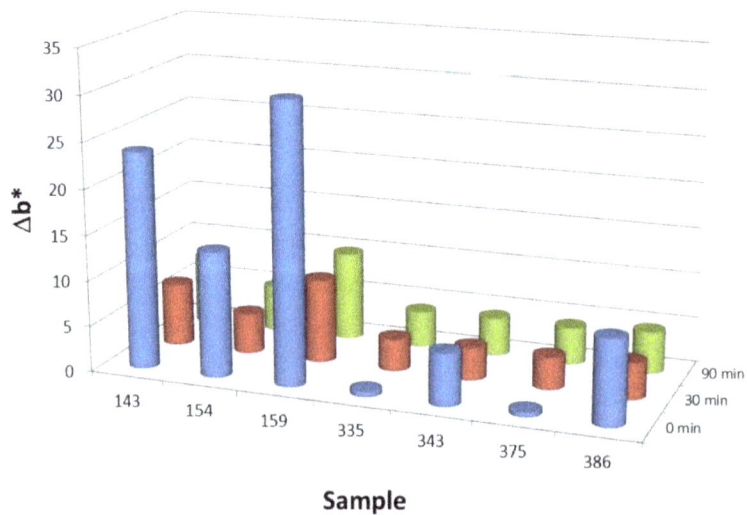

**Figure 11.** Changes of Δbx* for irradiated samples at different times.

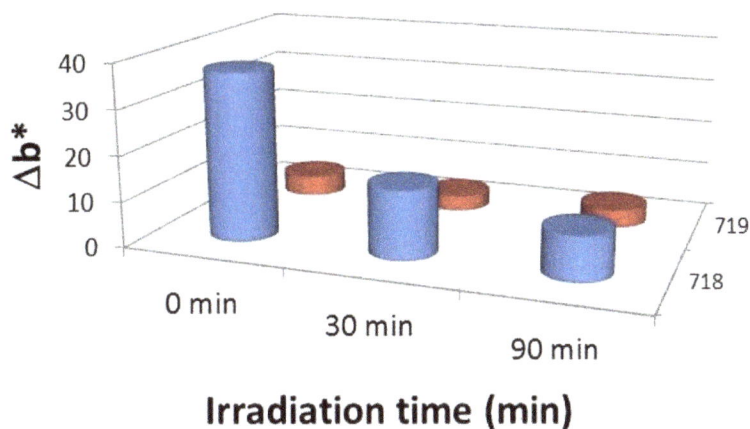

**Figure 12.** Changes of Δbx* for irradiated metallic samples at different times.

During the irradiation with polychromatic light, in the UV–Vis spectra of these pigments, either in solution or painted on cardboard, some minor changes evidenced in the spectral region between 480 and 840 nm could be observed, as follows:

- An insignificant decrease of Heliogen Blau 6900 absorbance, both at 660 and 860 nm, concomitant with a small increase of absorbance at 472 nm (**Figure 13**).

- A strong decrease of Heliogen Grün L8730 absorbance, visible both for Soret band from 447 nm and for the bands from 690 and 805 nm, most probably due to the high number of chlorine atoms present as substituents in the pigment structure, responsible for new generated species. The irradiation with Hg lamp (which contains a high level of UV radiation) causes dissociation of the *chlorine substituents*, leading to free radical formation [23] and new species visible by the new wideband from 495 nm (**Figure 14**).

- This behavior is exhibited for the cardboard painted with Heliogen Grün L8730, the spectral changes and absorbance decrease being observed most predominantly in the visible area (the absorption bands from 690 and 805 nm); in the near UV area, the changes attributed to

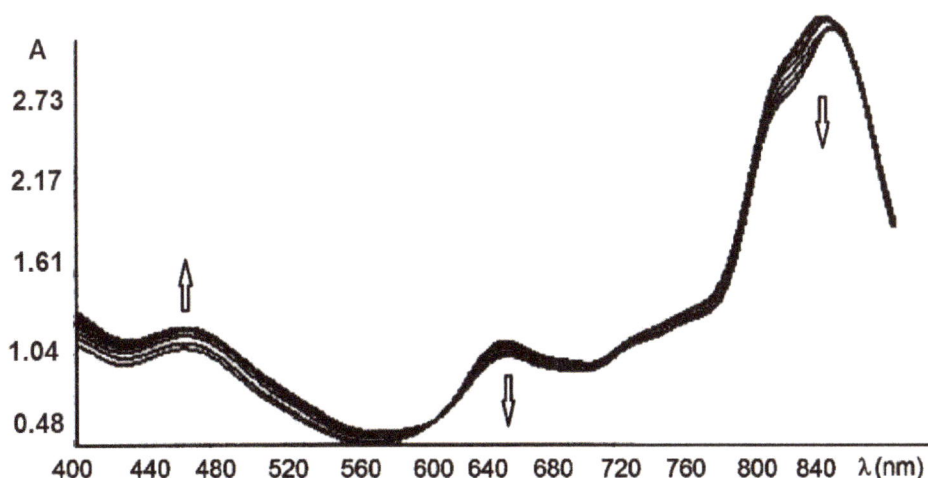

**Figure 13.** Spectral changes of Heliogen Blau 6900 in solution during light irradiation.

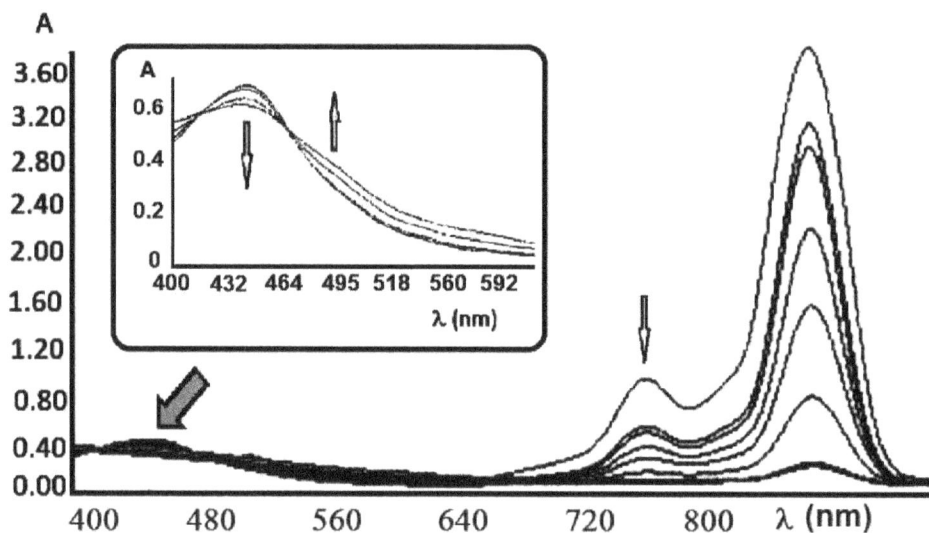

**Figure 14.** Spectral changes of Heliogen Grün L8730 in solution during light irradiation.

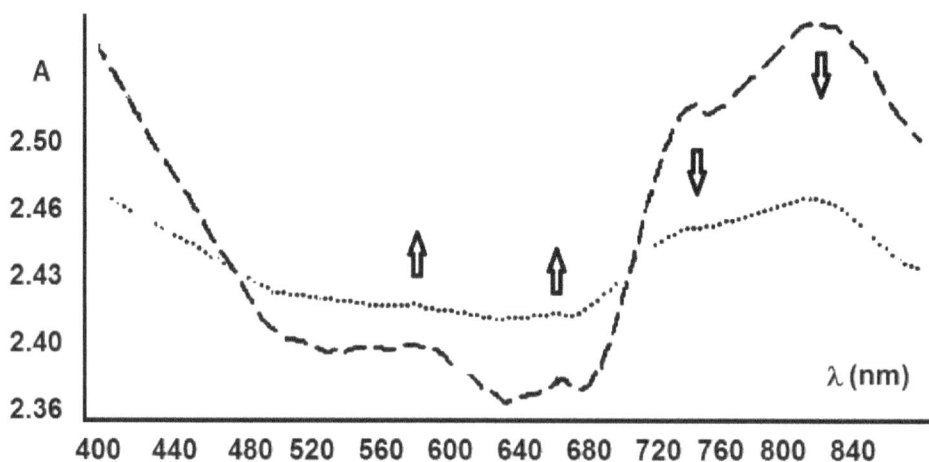

**Figure 15.** Spectral changes of cardboard painted with Phthalo Green during light irradiation.

Soret band are not visible, due to the presence of $TiO_2$ in the paint composition, this metallic oxide being responsible for the wide and not defined band from this area (**Figure 15**).

If the Pc oxidative photobleaching occurs via singlet oxygen species, the unsaturated bonds of phthalocyanine chromophoric system are oxidized [24], singlet oxygen being responsible for the phthalocyanine photobleaching [25]. There are some limited studies about photobleaching of MPcs in solution; because of its complex reaction, the product isolation is very difficult; and the final photooxidation products are obtained in insufficient amount (phthalimides and phthalonitriles) [26, 27]. Similar results have been described for porphyrin structures, which are related compounds with phthalocyanines [28–30]. Formation of highly toxic hydrogen cyanide upon ruby laser irradiation of the tattoo pigment Phthalocyanine Blue has been reported [31] (**Figure 16**).

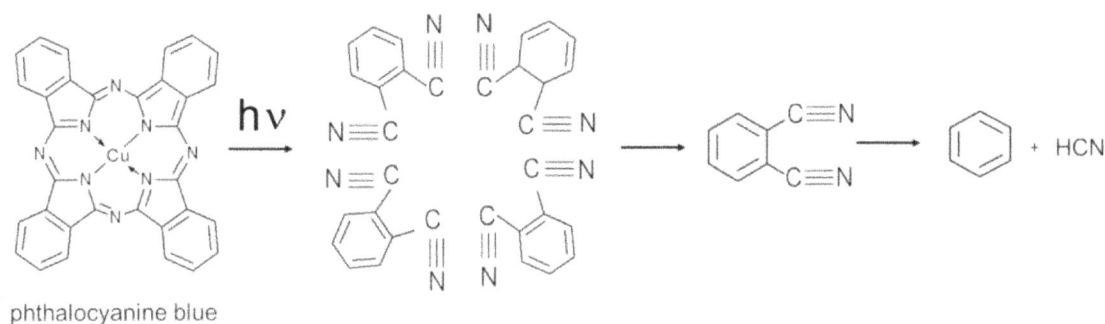

**Figure 16.** Decomposition pattern of Phthalocyanine Blue (PB15:3) based on the pyrolysis and laser irradiation.

## 5. Conclusions

By means of the abovementioned experiments, the following has been observed:

- In solution the green pigment is more degradable than the blue pigment.
- In acrylic paints containing these pigments, the photodegradation is more intense in the first 20–30 min of irradiation; after that, depending on the presence or absence of other compounds ($TiO_2$, micas, etc.), an increase of the optical parameters is observed.

Since the photodegradation results for pigments and acrylic paints are not fully correlated, some additional studies are needed in order to elucidate the photodegradation mechanism and products.

However, it is necessary to strictly observe that some protective measures regarding the uncontrolled exposure of paintings using these acrylic paints are necessary, in order to avoid the irreversible degradation processes of artworks.

## Acknowledgements

This paper received the financial support of the projects: PN 120BG/2016 and PN-III-P1-1.2-PCCDI2017-0476.

## Conflict of interest

The author(s) declared no potential conflicts of interest.

## Author informations

R. M. Ion, A. Nuta, A. A. Sorescu, and L. Iancu contributed equally to this work.

# Author details

Rodica-Mariana Ion[1,2]*, Alexandrina Nuta[1], Ana-Alexandra Sorescu[1] and Lorena Iancu[1,2]

* Address all correspondence to: rodica_ion2000@yahoo.co.uk

1 ICECHIM, Evaluation and Conservation of Cultural Heritage, Bucharest, Romania

2 Materials Engineering Doctoral School, Valahia University, Targoviste, Romania

# References

[1] Ion RM. Photochemistry. Principles and Applications. Bucharest: FMR Ed.; 2005

[2] Learner T. Modern Paints Uncovered. Los Angeles: The Getty Conservation Institute; 2008

[3] Jablonski E, Learner T, Hayes T, Golden M. The Conservation of Acrylic Emulsion Paintings: A Literature Review. Tate Papers, Tate's Online Research Journal; 2004

[4] Chiantore O, Trossarelli L, Lazzari M. Photooxidative degradation of acrylic and methacrylic polymers. Polymer. 2000;41:1657-1668

[5] Hamid H. Handbook of Polymer Degradation. New York: CRC Press; 2000

[6] Feller RL. Accelerated Ageing: Photochemical and Thermal Aspects. Los Angeles: The Getty Conservation Institute; 1994

[7] Pintus V, Ploeger R, Chiantore O, Wei S, Schreiner M. Thermal analysis of the interaction of inorganic pigments with p(nBA/MMA) acrylic emulsion before and after UV ageing. Journal of Thermal Analysis and Calorimetry. 2012;114:33-43

[8] Defeyt C, Strivay D. PB15 as 20th and 21st artists' Pigments: Conservation Concerns, e-PS. 2014;11:6-14

[9] Anghelone M, Jembrih-Simbürger D, Schreiner M. Identification of copper phthalocyanine blue polymorphs in unaged and aged paint systems by means of micro-Raman spectroscopy and random Forest. Spectrochimica Acta Part A. 2015;149:419-425. DOI: 10.1016/j.saa.2015.04.094

[10] Gettens RG, Stout GL. Pigment and Inert materials, In Painting Materials. A Short Encyclopaedia. D. Van Nostrand, New York; 1962. Dover Editions; 1966. pp. 91-185

[11] Moser FH, Thomas AL. Phthalocyanine Compounds. London: Chapman and Hall; 1963

[12] Quillen Lomax S. Phthalocyanine and quinacridone pigments: Their history, properties and use. Reviews in Conservation. 2005;6:19

[13] Mugarza A, Robles R, Krull C, Korytar R, Lorente N, Gambardella P. Electronic and magnetic properties of molecule-metal interfaces: Transition metal phthalocyanines adsorbed on ag(100). Physical Review B. 2012;85:55437. DOI: 10.1103/PhysRevB.85.155437

[14] Geuskens G, David C. The photo-oxidation of polymers – A comparison with low molecular weight compounds. Pure and Applied Chemistry. 1979;**51**:233-240

[15] Saron C, Zulli F, Giordano M, Felisberti MI. Influence of copper-phthalocyanine on the photodegradation of polycarbonate. Polymer Degradation and Stability. 2006;**91**(12): 3301-3311

[16] Rabek JF. Photostabilization of Polymers: Principles and Application. England: Elsevier Science Publisher Ltd; 1990

[17] Morlat S, Gardette J-L. Phototransformation of water-soluble polymers. I: Photo- and thermooxidation of poly(ethylene oxide) in solid state. Polymer. 2001;**42**:6071-6079

[18] de Sainte Claire P. Degradation of PEO in the solid state: A theoretical kinetic model. Macromolecules. 2009;**42**:3469e3482

[19] De Rosa MC, Crutchley RJ. Photosensitized singlet oxygen and its applications. Coordination Chemistry Reviews. 2002;**233/234**:351-371

[20] Spikes J, Lier J, Bommer JJ. A comparison of the photoproperties of zinc phthalocyanine and zinc naphthalocyanine tetrasulfonates: Model sensitizers for the photodynamic therapy of tumors. Photochemistry and Photobiology A. 1995;**91**:193-198

[21] d'Alessandro N, Tonucci L, Morvillo A, Dragani LK, Di Deo M, Bressan MJ, Thermal stability and photostability of water solutions of sulfophthalocyanines of Ru(II), Cu(II), Ni(II), Fe(III) and Co(II). Organometallic Chemistry. 2005;**690**:2133-2141

[22] Kuznetsova NA, Makarov DA, Yuzhakova OA, Solovieva LI, Kaliya OL. Study on the photostability of water-soluble Zn(II) and Al(III) phthalocyanines in aqueous solutions. Journal of Porphyrins and Phthalocyanines. 2010;**14**:968-974

[23] Moore DE. Drug-induced cutaneous photosensitivity: Incidence, mechanism, prevention and management. Drug Safety. 2002;**25**:345-337

[24] Bonnett R, Martinez G. Photobleaching of sensitizers used in photodynamic therapy. Tetrahedron. 2001;**57**:9513-9547. DOI: 10.1016/S0040-4020(01)00952-8

[25] Schnurpfeil G, Sobbi AK, Spiller W, Kliesch H, Wöhrle D. Photo-oxidative stability and its correlation with semi-empirical MO calculations of various tetraazaporphyrin derivatives in solution. Journal of Porphyrins and Phthalocyanines. 1997;**1**:159-167

[26] Cook MJ, Chambrier I, Cracknell SJ, Mayes DA, Russell DA. Octaalkyl zinc phthalocyanines: Potential photosensitizers for use in the photodynamic therapy of cancer. Photochemistry and Photobiology. 1995;**62**:542-545

[27] Kuznetsova NA, Kaliya OL. Oxidative photobleaching of phthalocyanines in solution. Journal of Porphyrins and Phthalocyanines. 2012;**16**:705-712. DOI: 10.1142/ S1088424612300042

[28] Ion RM. Spectrophotometric study of the photodegradation reaction of the tetra-aryl-porphyrins. The hydroperoxide effect. Revista de Chimie. 1995;**46**(2):134

[29]  Teodorescu L, Ion RM. New aspects on the photodegradation of the porphyrinic photo-sensitizers. Revista de Chimie. 1990;**41**(4):312-318

[30]  Siejak A, Wróbel D, Ion RM. Study of resonance effects in copper phthalocyanines. Journal of Photochemistry and Photobiology A: Chemistry. 2006;**181**(2-3):180-187

[31]  Schreiver I, Hutzler C, Laux P, Berlien HP, Luch A. Formation of highly toxic hydrogen cyanide upon ruby laser irradiation of the tattoo pigment phthalocyanine blue. Scientific Reports. 2015;**5**:12915. DOI: 10.1038/srep12915

# 4

# Fluorescence Dyes for Determination of Cyanide

Issah Yahaya and Zeynel Seferoglu

**Abstract**

Cyanides being highly poisonous to living beings and pollutants to our environment are among the most important anions studied over the years. As cyanide usage continues to sky-rocket, it is extremely important and high time that chemists devised methods for their detection to ensure harmless usage and safer working conditions for people coming into contact with cyanide and its compounds, day in day out. In this book, an attempt has been made to provide an in-depth commentary of literature for the synthesis of fluorescent dyes and mechanisms for the molecular recognition and detection of cyanide ions. It also covers some current entropy on colorimetric and fluorescent organic chemical probes for the detection and quantification of cyanide anions via fluorogenic and chromogenic procedures.

**Keywords:** fluorescent dyes, molecular recognition, fluorogenic, chromogenic procedures, cyanide detection, sensing mechanisms

## 1. Introduction

The design of protocol for selective optical signaling probes for anions has received much attention over the years as a result of the significant roles they play in biological and environmental procedures [1]. The recognition of cyanide has become an area of increasing significance in supramolecular chemistry as a result of the vital role it plays in environmental, clinical, chemical, and biological applications, and the fact that much attention has been given to the preparation of artificial probes that have the capability of uniquely recognizing and sensing anion species [2, 3]. Cyanide is famous for being one of the most toxic materials and is very dangerous to the environment and human health [4]. As a result of the extreme toxicity of cyanide ions in physiological [4–6] and environmental [7] systems, many investigators have designed optical probes [8, 9] for the sensitive and bias detection of cyanide. Till date, many strategies have been designed and developed for the detection of cyanide, including the formation of cyanide complexes with transition metal ions [10–14], boron derivatives [15, 16], CdSe quantum dots [17, 18], the displacement approach [19], hydrogen-bond interactions [20–22], deprotonation [23], and luminescence lifetime measurement [24]. For the interferences of competing anions to be curtailed in the sensing of cyanide, the nucleophilicity of the cyanide ion has been utilized, which includes its nucleophilic reactions with oxazine [25–27], pyrylium [28], squaraine [29], acyltriazene [19], acridinium [30], salicylaldehyde [31–33], trifluoroacetophenone [34–38], trifluoroacetamide derivatives [39–43], and other highly electrophilic carbonyl groups or imine [22, 44–48].

A lot of chemosensors for cyanide ion have been developed [49], chromogenic and fluorogenic probes for the detection of cyanide by the naked eye have attracted much interest as a result of the facile, fast usage, and their high sensitivity. As it is well-known, the probes are normally designed by the combination of a luminophore and an anion binding unit. Mostly, the anion binder is basically composed of H-bonding donors [50]. Herein, concise literature reports have been made on some strategies employed in the sensing of cyanide ions, dating from 2008 to 2017.

## 2. Sensing mechanisms and synthesis of cyanide sensors (CS)

The official methods of determining cyanides include titration [51, 52], spectrophotometry [65, 66], potentiometry with cyanide-selective electrodes [51, 53], flow injection (FI)-amperometry [54]. Analysis of cyanide in various matrices including water, soil, air, exhaled breath, food, and biological fluids (blood, urine, saliva, etc.), have been reviewed in official documents [55, 56], books [57] and journal articles [58, 59–61]. Quiet recently, Xu et al. [62] and Zelder and Mannel-Croise [63] have respectively written reviews on optical sensors and colorimetric measurement of cyanide. Herein, different sensing strategies have been discussed.

### 2.1. Cyanide sensing via aggregation induced emission (AIE)

This uncommon fluorescence phenomenon was perceived by Luo et al. [64] in 2001 via a solution of 1-methyl 1,2,3,4,5-pentaphenylsilole, and the term aggregation-induced emission was given to it. Tang et al. gave an explanation on the AIE phenomenon through a series of experimental analyses. They realized that the main cause of the AIE phenomenon was due to restriction of intramolecular rotation in the aggregates.

Sun et al. [65] have prepared a turn-on fluorescent probe **CS1** based on terthienyl for the detection of cyanide through the aggregation-induced emission (AIE) behavior of terthienyl units in aqueous solutions (**Scheme 1**). They confirmed the AIE behavior of **CS1** using the dynamic light scattering (DLS) measurements and the scanning electron microscopy (SEM) studies. The UV-vis titration of the free **CS1** showed absorption at 530 nm with a distinct color. Upon adding CN⁻, the absorption was quenched and eventually disappeared when the concentration of CN⁻ reached 40 μM. In the fluorescence studies of the solution of the probe, it showed non-emissiveness in the absence of CN⁻. Upon addition of cyanide to the solution

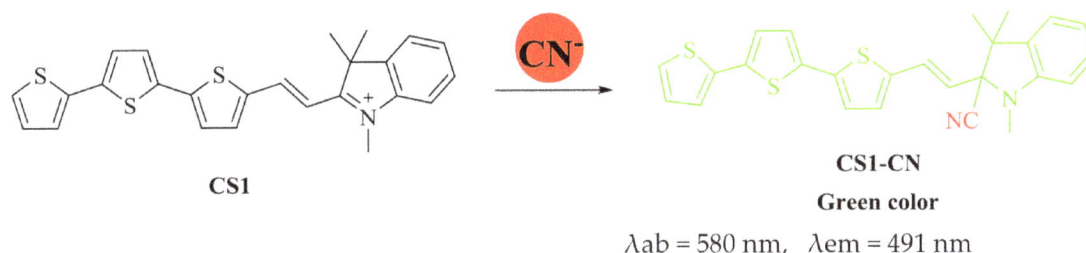

CS1

CS1-CN

**Green color**

$\lambda ab = 580$ nm,   $\lambda em = 491$ nm

**Scheme 1.** Sensing mechanism of cyanide employing **CS1**.

of **CS1**, an increase in its intensity at 491 nm was observed, and the fluorescence intensity further increased by the addition of more than 170 times. As the concentration of $CN^-$ reached 40 μM, a bright green fluorescent emission was observed which can easily be noticed by the naked eyes.

Another AIE probe **CS2** has been prepared by Chen et al. [66]. The sensor was synthesized using 2-benzothiazoleacetonitrile and 4-(diphenylamino)-benzaldehyde in 73% yield. The investi-gators synthesized the probe by placing equimolar (10 mmol) of 2-Benzothiazoleacetonitrile, ammonium acetate, and 4-(diphenylamino)benzaldehyde in ethanol (30 mL). The reaction was stirred at room temperature overnight. Then obtained product was filtered and recrystal-lized from dichloromethane (5 mL) and ethanol (50 mL) to afford the product. The absorption and fluorescence titrations of the probe (5 μM) were used to ascertain its AIE properties in acetonitrile/water (1: 99, v/v) solvent mixture at room temperature. The researchers employed different anions including $CN^-$, $HSO_4^-$, $SO_4^{2-}$, $HSO_3^-$, $CH_3COO^-$, $Cl^-$, $Br^-$, $I^-$, $F^-$, $NO_3^-$ and $H_2PO_4^-$ in the analysis. All the anions, with the exception of cyanide, exhibited almost no changes in the fluorescence intensity. However, the addition of cyanide led to 99% decrease in the fluorescence intensity of **CS2**, which confirmed that **CS2** could significantly sense cyanide. When **CS2** was dissolved in acetonitrile, a weakly fluorescent was seen. The authors found **CS2** to be non-emissive in acetonitrile. However, upon addition of large amounts of water (fw > 80 vol%) to acetonitrile, an orange fluorescence ($\lambda_{em}$ = 580 nm) was observed under identical measurement conditions (**Scheme 2**). However, the group found the absorbance of the solution to be weak as the fraction of water was below 80%. Additionally, the fluores-cence intensity with 90% water content was a bit intense than that with 99% which may be attributed to a more perfect aggregation state. Nonetheless, the characterizations were carried out in $CH_3CN$/water (1:99, v/v) (almost 100% aqueous solution) in consideration of practical use and environment protection. They therefore proposed cyanide sensing mechanism using probe **CS2** as illustrated in **Scheme 2**.

## 2.2. Cyanide sensing via the chemodosimeter approach

The special nucleophilic character of cyanide has been utilized for the preparation of different chemodosimetric sensors for cyanide, mostly in aqueous solutions.

In 2009, Kim and Kim [67] prepared, through the condensation reaction of nitromethane and a coumarinyl aldehyde, a new fluorescent chemodosimeter **CS3**, and investigated it for the detection of cyanide ions. The sensor has a coumarin moiety as the fluorescent signaling unit

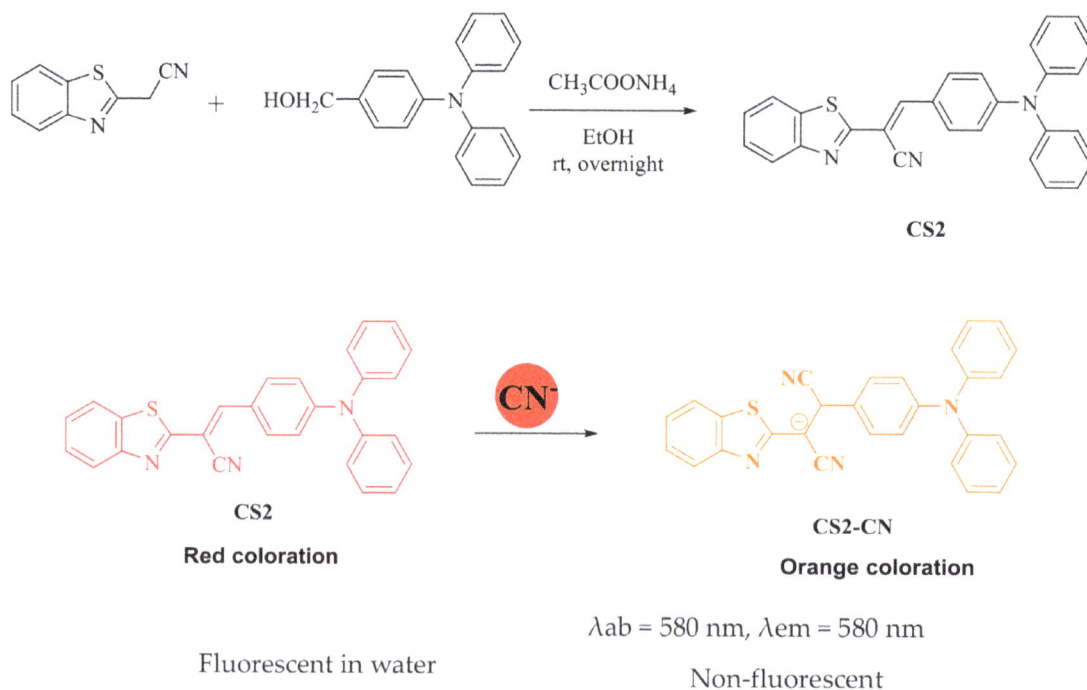

**Scheme 2.** Synthesis and sensing mechanism of probe **CS2**.

and a Michael acceptor unit being an unsaturated nitro group toward the cyanide. The authors proposed, using spectroscopic and chromatographic evidence, a plausible mechanism for the Michael acceptor type chemodosimeter **CS3**. Because the unsaturated nitro group of **CS3** is one of the good Michael acceptors, cyanide can be added to the β- or δ-position of the unsaturated nitro group, where the δ-position is doubly activated. The chemical reaction of **CS3** with a cyanide nucleophile was reported to be capable of causing a change in the electronic structure of the sensor thereby inducing a color change from light orange to pink at 468 nm (**Scheme 3**).

Hu et al. [68] have successfully synthesized and reported a 1,3-indanedione-based chemodosimeter that could be employed in sensing cyanide ions via both aggregation-induced emission enhancement (AIEE) and intramolecular charge transfer (ICT) in 90% aqueous medium. They prepared a solution of the chemodosimeter ($1.0 \times 10^{-5}$ M) in aqueous solution (THF:$H_2O$ = 1:9 [v/v], containing $10 \times 10^{-3}$ M HEPES, pH = 7.3). In the aqueous medium, the chemodosimeter showed a strong ICT absorption band at 425 nm, and upon adding CN⁻, the ICT band was said to have disappeared and the color of the solution changed from yellow to colorless.

## 2.3. Cyanide sensing via the excited state intra- and inter-molecular proton transfer (ESIPT)

In 2017, Huo and co-workers [69] reported the synthesis of a novel isophorone-based red-emitting fluorescent probe **CS4** that can be used in signaling cyanide ion through hampering of its ESIPT (**Scheme 4**). With the addition of cyanide ions, as reported by the investigators, the absorption spectrum of **CS4** (5 mM) exhibited an obvious peak at 419 nm. Upon further

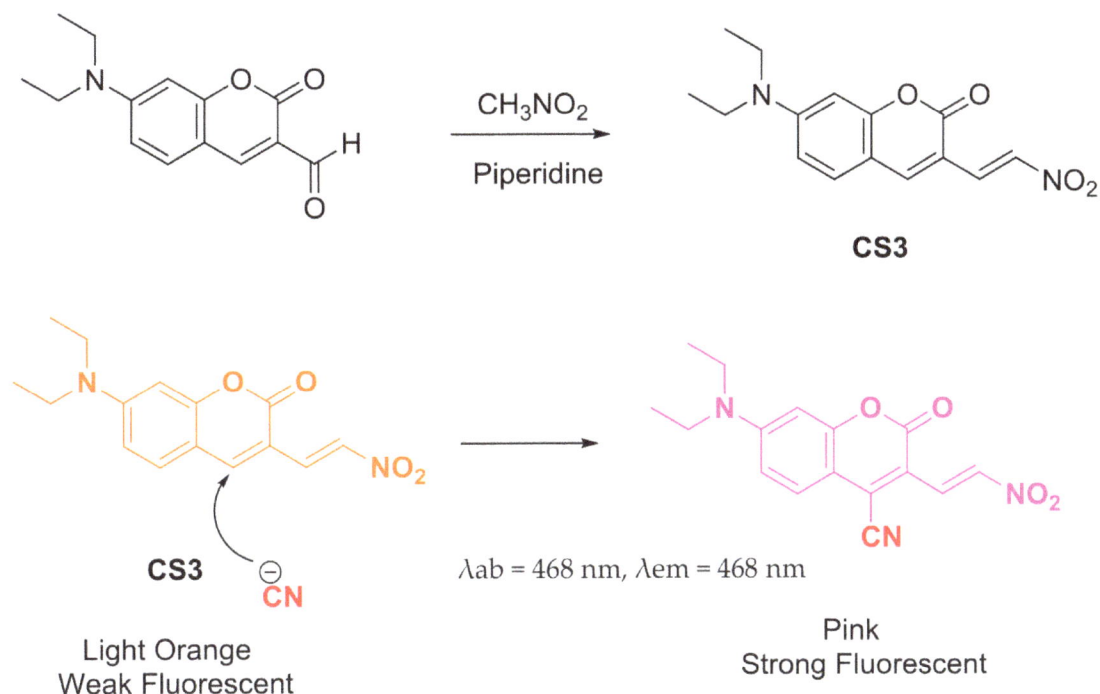

Scheme 3. Synthesis and plausible mechanism for cyanide using CS3.

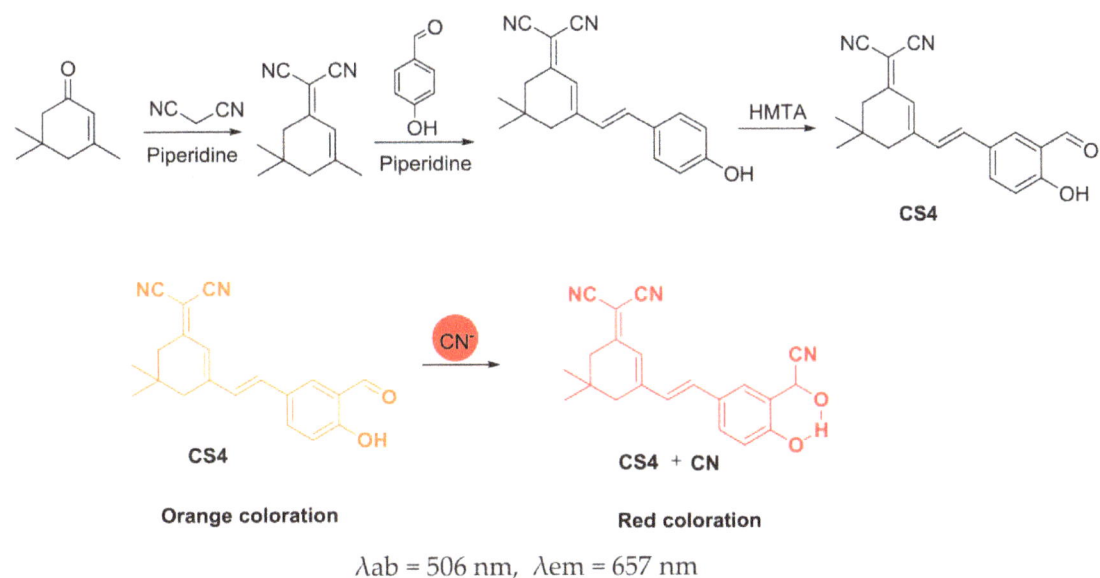

Scheme 4. Sensing mechanism for cyanide using CS4.

addition of CN⁻, the absorption band gradually weakened followed by a rapid step-up in the peak at 506 nm, intimating that **CS4** has involved in a nucleophilic reaction with the CN⁻, which led to a distinct color change from orange to red. The authors observed the fluorescence spectra of **CS4** (5 mM) without CN⁻ to have shown an emission at 616 nm with orange color under a handheld UV lamp. After adding CN⁻, the emission at 616 nm was said to have attenuated sharply and then followed by a peak increased at 657 nm, which induced a fluorescence chromogenic change from orange to red (**Scheme 4**).

Shymaprosad Goswami and co-researchers [70] have reported an ESIPT exhibiting benzothiazole receptor possessing two aldehyde groups; one ortho and the other para to an OH group. The ortho aldehyde group being very reactive, was reported to have undergone a nucleophilic reaction with CN⁻ selectively, thereby hampering an ESIPT. The investigators confirmed the process via DFT and TD-DFT computations. The affinity of the benzothiazole receptor toward different competing ions was investigated using UV-vis absorption and emission spectrometry in aqueous acetonitrile solution. The probe showed a green emission at 521 nm, a peculiar benzothiazolyl phenol ring emission. Upon adding CN⁻, the emission was reported to have drastically decreased, followed by an increase at 436 nm. This suggested that, as thought by the authors, a chemical reaction between the cyanide and the aldehyde group has interrupted the conjugation and thereby hampering the ESIPT process leading to a color change from green to blue. In the UV-vis absorption study, they found that, only CN⁻ had induced the perturbation of the electronic behavior of benzothiazole receptor.

## 2.4. Cyanide sensing via the excimer/exciplex form

Wang et al. [71] have successfully designed and reported the synthesis of a novel probe **CS5** comprising diketopyrrolopyrrole and indanedione-based Michael receptor. The sensor could be employed in the recognition of cyanide anion. The researchers realized that, as an aqueous solution of cyanide was added to **CS5** in THF, it induced a sudden color change from purple to yellow, as well as a large blue shift from 553 to 480 nm, without any interference from the other interfering ions ($F^-$, $Cl^-$, $Br^-$, $I^-$, $H_2PO_4^-$, $SO_4^{2-}$, $CO_3^{2-}$, $PO_4^{3-}$, and $OAc^-$). They also studied the binding of CN⁻ with **CS5**. Therein, they found out that, the absorption spectra of **CS5** in THF changed upon addition of an aqueous solution of cyanide. This led to absorption peaks of **CS5** at 359 and 553 nm, which gradually attenuated following the formation of two new bands centered at 314 and 480 nm with color change from purple to yellow (**Scheme 5**).

Shahid et al. [72] have prepared and described a new simple organic scaffold based on acenaphthene. The fluorogenic and chromogenic properties of the probe were investigated for signaling metal cations and anions in $H_2O/CH_3CN$ (8:2, v/v) solvent mixture. The authors employed a metal chelate based sensing strategy of copper complexes for fluorescent sensing of cyanide.

## 2.5. Cyanide sensing via the Förster/fluorescence resonance energy transfer (FRET)

Goswami et al. [73] investigated a chemosensor **CS6** having a naphthalene and fluorescein chromophores acting as an energy donor and an acceptor, respectively. The authors reported that, the emission of the naphthalene chromophore and the absorption of the ring-opened fluorescein dye indicated that, there was an overlap between these two spectra, implying that the FRET from the naphthalene chromophore to the fluorescein moiety had occurred. Additionally, they found out that, the probe **CS6**-$Zn^{2+}$ complex for signaling cyanide was generated *in situ* by adding 1 equiv of $ZnCl_2$ to $CH_3OH/H_2O$ (3:7, v/v) solution of probe **CS6**. They conducted the UV-vis and fluorescence studies at pH 7.1, and realized that there was the disappearance of yellow color of the **CS6**-$Zn^{2+}$ with increased concentration of CN⁻. The researchers suggested that, the observed decrease of yellow coloration of the solution

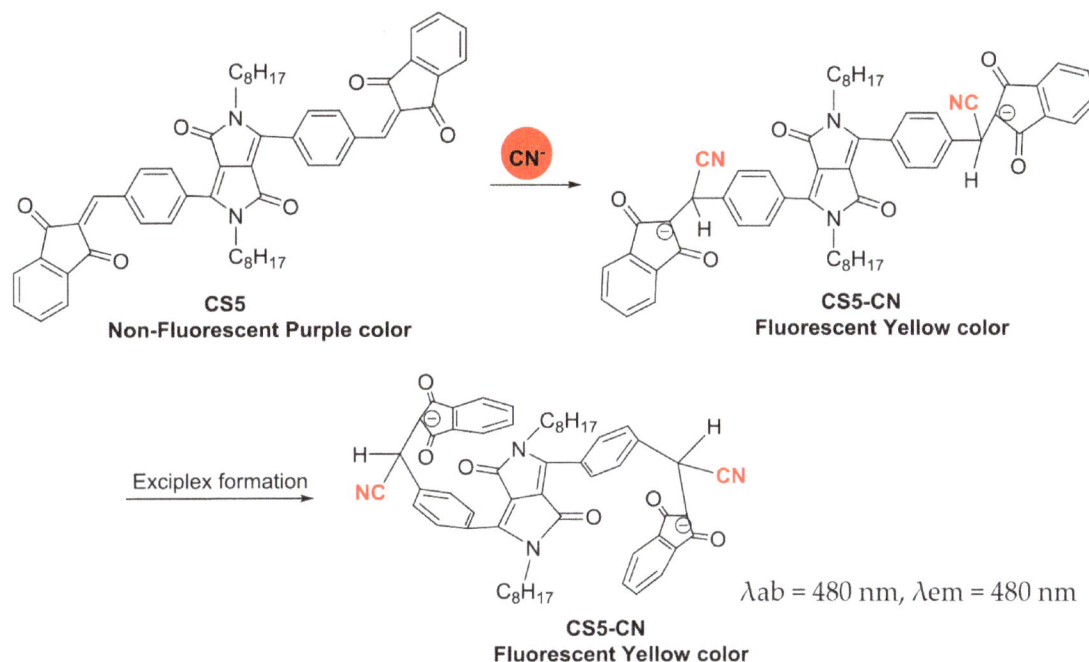

**Scheme 5.** Structure of **CS5** and its cyanide sensing.

containing the **CS6**-Zn$^{2+}$ complex was a result of the ring opened amide form of **CS6**-Zn$^{2+}$ has been converted to the spirolactam form of **CS6** in the presence of CN$^-$. Again, the authors investigated the selectivity of cyanide through the UV response of **CS6**-Zn$^{2+}$ in the presence of competing anions such as Br$^-$, Cl$^-$, I$^-$, F$^-$, ADP, ATP, PPi, OAc$^-$, Pi, SH$^-$, SCN$^-$, and N$_3^-$. As cyanide solution was added to the solution of the **CS6**-Zn$^{2+}$ complex, there was a reverse change in the fluorescence spectra due to the occurrence of a reverse FRET phenomenon (**Scheme 6**).

In 2009, Chung and co-workers [74] successfully developed a cyanide sensor for fluorescence study. In the fluorescence study, different anions, such as CN$^-$, SCN$^-$, AcO$^-$, F$^-$, Cl$^-$, Br$^-$, I$^-$, H$_2$PO$_4^-$, HSO$_4^-$, NO$_3^-$, and ClO$_4^-$ were evaluated at pH 7.4 (0.02 M pH 7.4 HEPES). Using 100 equiv. of each of these anions, and 6 mM of the probe in the presence of Cu$^{2+}$ (1 equiv.), only CN$^-$ was observed to have shown a large fluorescence enhancement.

### 2.6. Cyanide sensing via H-bonding

In 2015, our group [75] designed and reported the synthesis and application of chemosensor **CS7** as cyanide sensor. The signaling performance of **CS7** was investigated using UV-vis absorption and fluorescence spectroscopy in DMSO against F$^-$, Cl$^-$, Br$^-$, I$^-$, AcO$^-$, CN$^-$, HSO$_4^-$ and ClO$_4^-$ anions with tetrabutylammonium (TBA) as the counter cation. The absorption maximum of **CS7** was observed at 405 nm in DMSO. This absorption band shifted hypsochromically upon addition of one equiv of CN$^-$ to the solution of the probe, and a new band at 392 nm was observed. Apparently, the electrostatic interaction between probe **CS7** and CN$^-$ deprotonated the amide -NH function and negative charge accumulated around the amide function giving rise to the observed hypsochromic shift. To gain, an insight into the binding of **CS7** with anions and fluorescence titrations were performed

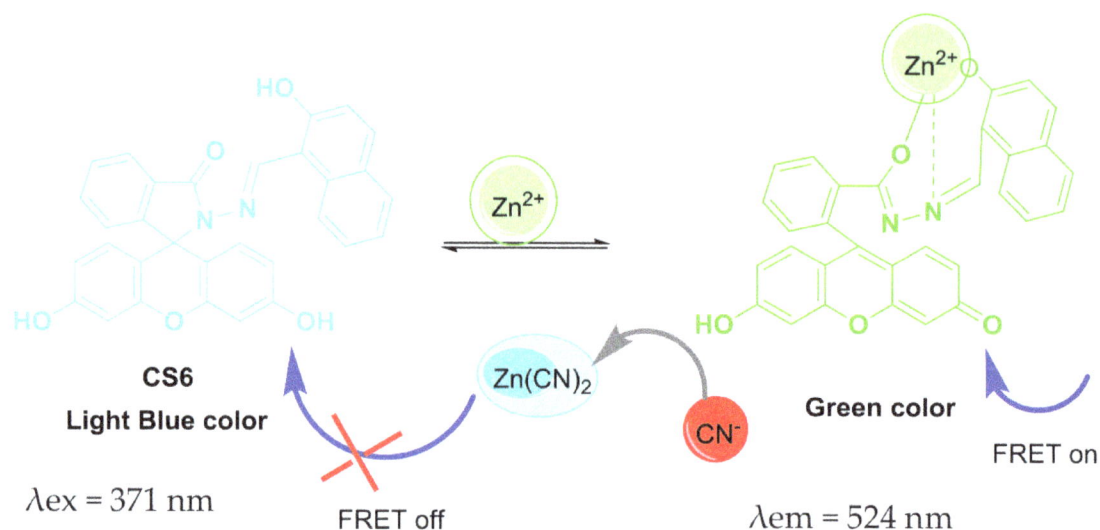

**Scheme 6.** Structure of **CS6** and cyanide sensing on the via a Zn-complex.

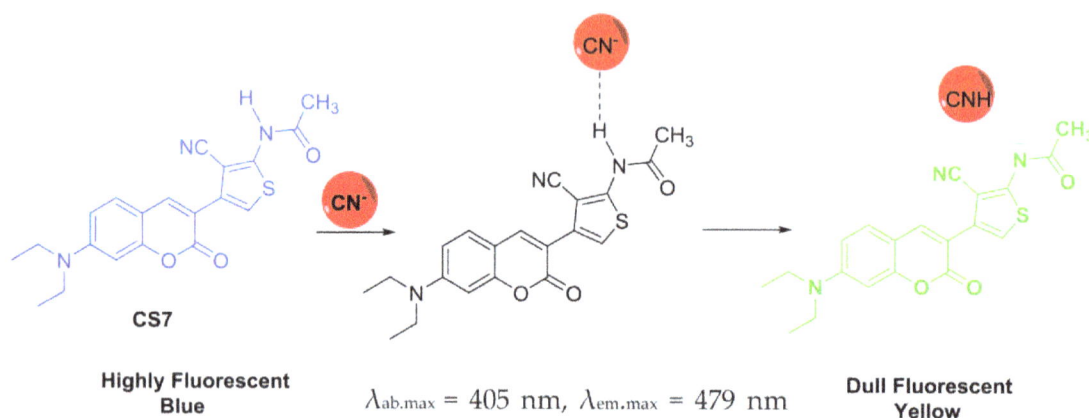

**Scheme 7.** Binding mode of **CS7** for cyanide sensing.

in DMSO with excitation wavelength of 405 nm. Free sensor, **CS7** displayed an emission maximum at 479 nm in DMSO with a high intensity. Upon addition of CN⁻ (>1 equiv) to the solution of **CS7**, the fluorescence intensity gradually decreased and emission color changed. Addition of 20 equiv of CN⁻ almost wiped out the fluorescence of **CS7**. Similar results were observed for the competing anions (**Scheme 7**).

A group of researchers [76] reported the synthesis of two receptors of specific signaling of cyanide ions in sodium cyanide solution. The authors associated the visual detection of CN⁻ via color changes, with the formation of hydrogen bonded adducts. They found the probes to have limited solubility in water, and therefore employed mixed solvent, such as $CH_3CN$/ HEPES buffer (1:1, v/v), for the sensing studies. The fluorogenic and visually detectable chromogenic changes of the receptors were verified using aqueous solutions of the sodium salt of all the employed common anionic analytes such as F⁻, Cl⁻, Br⁻, I⁻, CN⁻, SCN⁻, $CH_3COO^-$, $H_2PO_4^-$, $P_2O_7^{3-}$, $HSO_4^-$, $NO_3^-$, and $NO_2^-$ present in excess (0.1 mM). For the contending anions, no spectral changes in their spectral patterns was observed by the investigators. However, the researchers observed changes in spectral pattern, naked-eye color, and fluorescence, only in the presence of added CN⁻. Interference studies conducted by them revealed that, the spectral response for CN⁻ remained unaffected in the presence of 10 equiv of all interfering anions.

## 2.7. Cyanide sensing via the inter- or intra- molecular charge transfer (ICT)

In 2017, Hao et al. [77] designed and synthesized a probe **CS8** having a naphthalimide unit as the fluorophore and a methylated trifluroacetamide moiety as the acceptor part, which can be employed for selective and ratiometric signaling of cyanide. The group also employed the probe to study the sensing recognition of cyanide using HPLC, UV-vis, and emission spectroscopic analyses. The regioselective nucleophilic attack of cyanide ion to the methylated trifluroacetamide unit in the sensor was reported to induce an enhanced ICT process, and therefore causing a sudden red shift in both absorption and emission spectra of the sensor at 450 nm and 535 nm, respectively (**Scheme 8**).

Mashraqui and co-workers [78] have developed a novel chemodosimeter that has the structural capabilities to convert the CN-binding event into an enhanced ICT process, inducing absorbance red shifts and a high fluorescence turn-on response. The use of the probe toward sensing different anions was investigated by the group via optical spectral analysis. The group realized that, the absorption spectra of the chemosensor (28 μM) in DMSO-$H_2O$ (7:3, v/v) in tris-HCl buffer pH 7.0, was insensitive to each of the competing anions (F⁻, AcO⁻, SCN⁻, $HSO4⁻$, $NO_3⁻$, Br⁻, Cl⁻, I⁻, and $H_2PO_4⁻$) up to 75 mM. On the contrary, the concentration of cyanide (7.6 mM) which was at a 10-fold lower that of the interfering anions, was noticed to have elicited a monumental interaction, which was followed by an instant color change from colorless to deep yellow, an event that allowed selective visual detection of cyanide by the naked eye.

## 2.8. Cyanide sensing via the nucleophilic approach

Kwon et al. [79] successfully introduced a fluorescent chemodosimeter **CS9** which exhibited a green fluorescence coloration upon adding cyanide ions. The investigators analyzed the probe in aqueous solution and noticed that the probe showed an 'OFF-ON' type of emission change which could be utilized as a monitoring device for cyanide over 500 nm (**Scheme 9**).

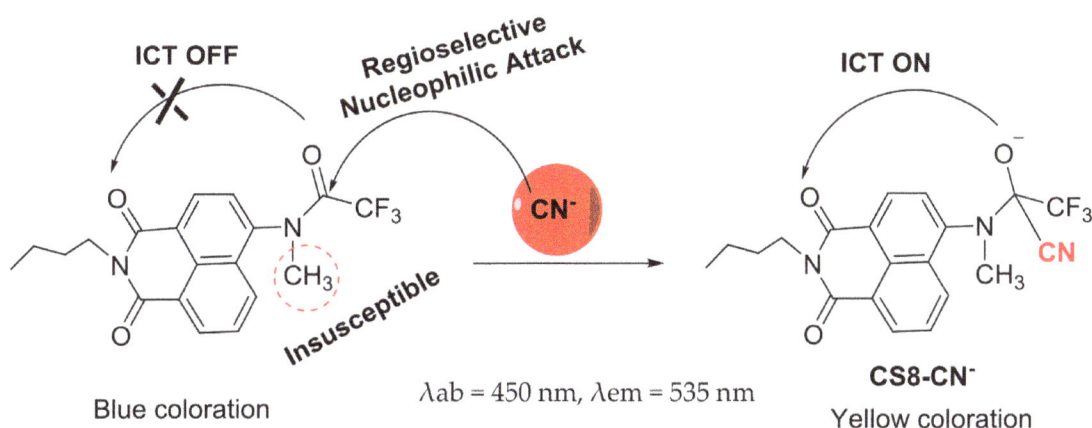

**Scheme 8.** Mechanism for the signaling of **CS8** with cyanide.

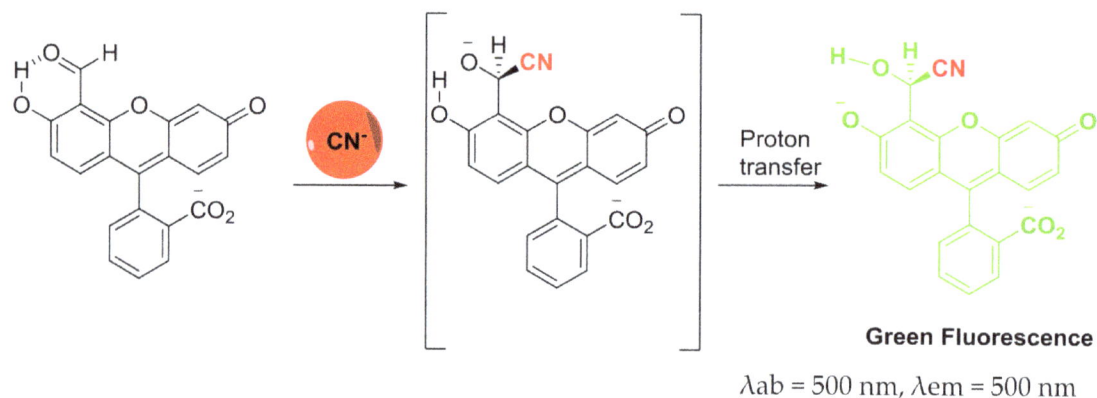

**Green Fluorescence**

$\lambda ab$ = 500 nm, $\lambda em$ = 500 nm

**Scheme 9.** Signaling of **CS9** upon addition of cyanide.

For the fluorescent analysis, 100 equiv of different anions ($CN^-$, $AcO^-$, $F^-$, $Cl^-$, $Br^-$, $I^-$, $H_2PO_4^-$, $HSO_4^-$, $NO_3^-$, and $ClO_4^-$) were investigated in acetonitrile—HEPES (9:1, v/v, 0.01 M pH 7.4 HEPES), containing probe **CS9** (3 μM). The authors attributed the selective recognition of cyanide as a result of the high nucleophilic nature of $CN^-$ in aqueous solution. Finally, they further studied the practical application of the sensor by applying it for the selective detection of cyanide in the living cells.

Li et al. [80] reported the development of selective and sensitive red-emitting fluorogenic and colorimetric dual-channel sensor for detection of cyanide. The group realized that, adding cyanide ion to the probe led to the display of huge blue-shift in both fluorescence (130 nm) and absorption (100 nm) spectra. The authors found that, the probe could be capable of selective signaling of cyanide by the naked-eye. They therefore concluded that, the mechanism for the detection of cyanide was due to the nucleophilic attack of cyanide toward the benzothiazole group of probe, which could block conjugation between benzothiazole unit and the naphthopyran moiety, resulting in both color and spectral changes.

$\lambda ab$ = 500 nm, $\lambda em$ = 500 nm.

## 2.9. Cyanide sensing via the photoinduced electron transfer (PET)

Qu et al. [81] described the synthesis of a fluorescent and colorimetric chemosensor **CS10** derived from a naphtho[2,1-*b*]furan-2-carbohydrazide and 2-hydroxy-1-naphthaldehyde, through a straightforward reaction, from inexpensive reagents, for a swift signaling, superior selectivity, and superb sensitivity toward cyanide ions. The researchers observed that, the mechanism for the recognition of cyanide ions was as a result of the photo-induced electron transfer (PET) (**Scheme 10**). The investigators found the probe **CS10** to possess a strong anti-interference toward other common anions ($F^-$, $AcO^-$, $H_2PO_4^-$, and $SCN^-$). The authors applied the sensor for detection of $CN^-$ in food samples, which they found to be an easier and selective platform for on-site monitoring of $CN^-$ in agriculture samples. The investigators carried out both UV-vis and the fluorescence spectroscopy experiments in EtOH/$H_2O$ (7:3,v/v) HEPES solution of sensor **CS10**. A significant color change from colorless to yellow, which was visible to the naked eyes, and it was accompanied by a strong and broad absorption red shift from 373 nm to 477 nm in the UV-visible spectrum of solution of the sensor in EtOH/$H_2O$ (7:3, v/v) HEPES solution. When

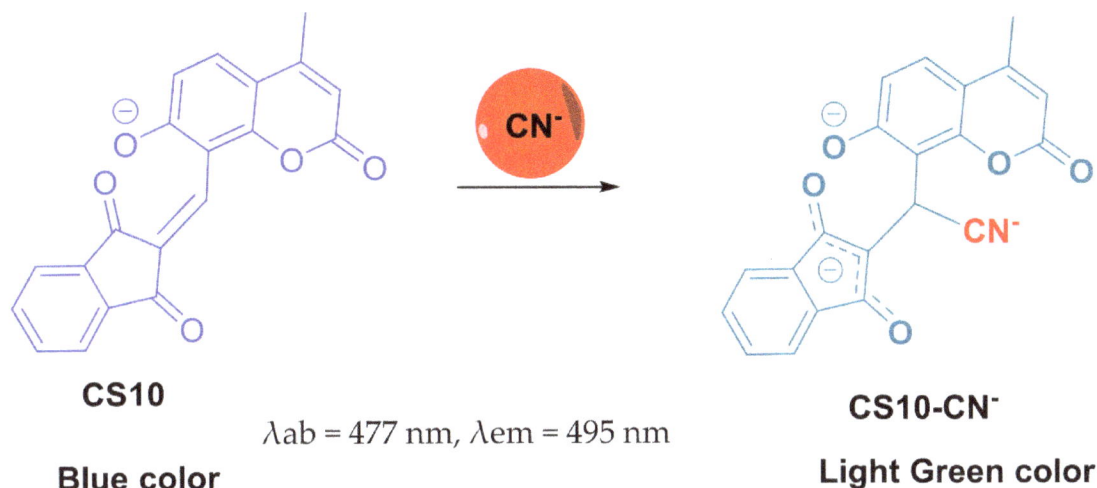

**CS10**

**Blue color**

$\lambda ab = 477$ nm, $\lambda em = 495$ nm

**CS10-CN⁻**

**Light Green color**

**Scheme 10.** Mechanism for the recognition of cyanide using sensor **CS10**.

50 equivalents of CN⁻ was added to the solution of the probe, the fluorescence intensity of the sensor **CS10** increased rapidly and the observed change in coloration from dim blue to blue-green was said to be distinguishable by the naked eye under the UV-lamp.

A group of researchers [82] successfully prepared a Co(II)-salen based fluorescent sensor that is applicable for selective recognition of cyanide anions in 1:2 binding stoichiometry. The scientists related the fluorescence enhancement of the solution of the probe, upon the addition of cyanide, to an interruption of photoinduced electron transfer from the coumarin fluorophore of the sensor to the cobalt(II) ion. In order to address the origin of the fluorescence enhancement of the sensor by the coordination of cyanide anions, the authors measured the HOMO and LUMO energy levels of the cobalt-salen complex of the chemosensor in the absence and the presence of cyanide anions via cyclic voltammetric and UV-vis spectroscopic measurements.

# 3. Conclusion

In summary, this chapter is limited to literature reports that have been published from 2008 to 2017. Some papers that have been published pre-2008 may have been used to illustrate important points. Some of the schemes for the synthetic pathways of the reported literature have not been illustrated in this chapter due to the limited space available to the authors. There are also a few papers that have been published within the period captured herein but could not be included. The omission of such literature does not in any way connote that such papers are of lesser importance.

# Acknowledgements

We appreciate, with kind regards, the support and sponsorship of this endeavor by Gazi University and TUBITAK for the grant (Grant numbers; 111 T106, 113Z704, 114Z980, and 215Z567).

## Conflict of interest

There is no conflict of interest, whatsoever, in publishing this piece.

## Author details

Issah Yahaya and Zeynel Seferoglu*

*Address all correspondence to: znseferoglu@gazi.edu.tr

Department of Chemistry, Gazi University, Ankara, Turkey

## References

[1] (a) Lee C-H, Miyaji H, Yoon D-W, Sessler JL. Strapped and other topographically non-planar calixpyrrole analogues. Improved anion receptors. Chemical Communications. 2008;(1):24-34 (b) Yoon J, Kim SK, Singh NJ, Kim KS. Imidazolium receptors for the recognition of anions. Chemical Society Reviews. 2006;**35**(4):355-360 (c) Gunnlaugsson T, Glynn M, Tocci GM, Kruger PE, Pfeffer FM. Anion recognition and sensing in organic and aqueous media using luminescent and colorimetric sensors. Coordination Chemistry Reviews. 2006;**250**(23-24):3094-3117 (d) Gale PA. From anion receptors to transporters. Accounts of Chemical Research. 2011;**44**(3):216-226 (e) Koskela SJM, Fyles TM, James TD. A ditopic fluorescent sensor for potassium fluoride. Chemical Communications. 2005;(7):945-947 (f) Kim SK, Kim HN, Xiaoru Z, Lee HN, Lee HN, Soh JH, Swamy KMK, Yoon J. Recent development of anion selective fluorescent chemosensors. Supramolecular Chemistry. 2007;**19**(4-5):221-227 (g) Martínez-Máñez R, Sancanón F. Fluorogenic and chromogenic chemosensors and reagents for anions. Chemical Reviews. 2003;**103**(11):4419-4476 (h) Czarnik AW. Chemical communication in water using fluorescent chemosensors. Accounts of Chemical Research. 1994;**27**(10):302-308 (i) Lee HN, Xu Z, Kim SK, Swamy KMK, Kim Y, Kim S-J, Yoon J. Pyrophosphate-selective fluorescent chemosensor at physiological pH: formation of a unique excimer upon addition of pyrophosphate. Journal of the American Chemical Society. 2007;**129**(13):3828-3829

[2] Beer PD, Gale PA. Anion recognition and sensing: The state of the art and future perspectives. Angewandte Chemie, International Edition. 2001;**40**(3):486-516

[3] Amendola V, Esteban-Gómez D, Fabbrizzi L, Licchelli M. What Anions do to N−H-containing receptors. Accounts of Chemical Research. 2006;**39**(5):343-353

[4] Bianchi A, Bowman-James K, Garcia-Espana E. Supramolecular Chemistry of Anions. New York, NY, USA: Wiley-VCH; 1997

[5] Timofeyenko YG, Rosentreter JJ, Mayo S. Piezoelectric quartz crystal microbalance sensor for trace aqueous cyanide ion determination. Analytical Chemistry. 2007;**79**(1):251-255

[6] Vennesland B, Comm EE, Knownles CJ, Westly J, Wissing F. Cyanide in biology. London: Academic; 1981

[7] Young C, Tidwell L, Anderson C. Cyanide: Social, Industrial, and Economic Aspects, Minerals, Metals, and Materials Society. Warrendale; 2001

[8] Xu Z, Chen X, Kim HN, Yoon J. Sensors for the optical detection of cyanide ion. Chemical Society Reviews. 2010;**39**(1):127-137

[9] Cho DG, Sessler JL. Modern reaction-based indicator systems. Chemical Society Reviews. 2009;**38**(6):1647-1662

[10] Kim YH, Hong JI. Ion pair recognition by Zn–porphyrin/crown ether conjugates: Visible sensing of sodium cyanide. Chemical Communications. 2002;(5):512-513

[11] Liu H, Shao XB, Jia MX, Jiang XK, Li ZT, Chen GJ. Selective recognition of sodium cyanide and potassium cyanide by diaza-crown ether-capped Zn-porphyrin receptors in polar solvents. Tetrahedron. 2005;**61**(34):8095-8100

[12] Chow CF, Lam MH, Wong WY. A heterobimetallic ruthenium (ii)–copper (ii) donor–acceptor complex as a chemodosimetric ensemble for selective cyanide detection. Inorganic Chemistry. 2004;**43**(26):8387-8393

[13] Zelder FH. Specific colorimetric detection of cyanide triggered by a conformational switch in vitamin B12. Inorganic Chemistry. 2008;**47**(4):1264-1266

[14] Zeng Q, Cai P, Li Z, Qin J, Tang BZ. An imidazole-functionalized polyacetylene: Convenient synthesis and selective chemosensor for metal ions and cyanide. Chemical Communications. 2008;(9):1094-1096

[15] Badugu R, Lakowicz JR, Geddes CD. Enhanced fluorescence cyanide detection at physiologically lethal levels: Reduced ICT-based signal transduction. Journal of the American Chemical Society. 2005;**127**(10):3635-3641. DOI: 10.1021/ja044421i

[16] Ros-Lis JV, Martínez-Máñez R, Sato J. A selective chromogenic reagent for cyanide determination. Chemical Communications. 2002;(19):2248-2249

[17] Jin WJ, Fernández-Argüelles MT, Costa-Fernández JM, Pereiro R, Sanz-Medel A. Photoactivated luminescent CdSe quantum dots as sensitive cyanide probes in aqueous solutions. Chemical Communications. 2005;(7):883-885

[18] Touceda-Varela A, Stevenson EI, Galve-Gasion JA, Dryden DT, Mareque-Rivas JC. Selective turn-on fluorescence detection of cyanide in water using hydrophobic CdSe quantum dots. Chemical Communications. 2008;(17):1998-2000

[19] Chung Y, Ahn KH. N-acyl triazenes as tunable and selective chemodosimeters toward cyanide ion. The Journal of Organic Chemistry. 2006;**71**(25):9470-9474

[20] Sun SS, Lees AJ. Anion recognition through hydrogen bonding: a simple, yet highly sensitive, luminescent metal-complex receptor. Chemical Communications. 2000;(17): 1687-1688

[21] Miyaji H, Sessler JL. Off-the-shelf colorimetric anion sensors. Angewandte Chemie, International Edition. 2001;**113**(1):158-161

[22] Saha S, Ghosh A, Mahato P, Mishra S, Mishra SK, Suresh E, Das S, Das A. Specific recognition and sensing of CN− in sodium cyanide solution. Organic Letters. 2010;**12**(15): 3406-3409

[23] Gimeno N, Li X, Durrant JR, Vilar R. Cyanide sensing with organic dyes: Studies in solution and on nanostructured Al$_2$O$_3$ surfaces. Chemistry - A European Journal. 2008;**14**(10):3006-3012

[24] Anzenbacher P, Tyson DS, Jursíková K, Castellano FN. Luminescence lifetime-based sensor for cyanide and related anions. Journal of the American Chemical Society. 2002;**124**(22):6232-6233

[25] Tomasulo M, Sortino S, White AJ, Raymo FM. Chromogenic oxazines for cyanide detection. The Journal of Organic Chemistry. 2006;**71**(2):744-753. DOI: 10.1021/jo052096r

[26] Ren J, Zhu W, Tian H. A highly sensitive and selective chemosensor for cyanide. Talanta. 2008;**75**(3):760-764

[27] Tomasulo M, Raymo FM. Colorimetric detection of cyanide with a chromogenic oxazine. Organic Letters. 2005;**7**(21):4633-4636. DOI: 10.1021/ol051750m

[28] Garcia F, Garcia JM, Garcia-Acosta B, Martínez-Máñez R, Sancenón F, Soto J. Pyrylium-containing polymers as sensory materials for the colorimetric sensing of cyanide in water. Chemical Communications. 2005;(22):2790-2792

[29] Ros-Lis JV, Martínez-Máñez R, Sato J. A selective chromogenic reagent for cyanide determination. Chemical Communications. 2002;(19):2248-2249

[30] Yang Y, Tae J. Acridinium salt based fluorescent and colorimetric chemosensor for the detection of cyanide in water. Organic Letters. 2006;**8**(25):5721-5723. DOI: 10.1021/ol062323r

[31] Lee KS, Kim HJ, Kim GH, Shin I, Hong JI. Fluorescent chemodosimeter for selective detection of cyanide in water. Organic letters. 2008;**10**(1):49-51

[32] Kwon SK, Kou S, Kim HN, Chen X, Hwang H, Nam SW, Yoon J. Sensing cyanide ion via fluorescent change and its application to the microfluidic system. Tetrahedron Letters. 2008;**49**(26):4102-4105

[33] Lee KS, Lee JT, Hong JI, Kim HJ. Visual detection of cyanide through intramolecular hydrogen bond. Chemistry Letters. 2007;**36**(6):816-817

[34] Kim YK, Lee YH, Lee HY, Kim MK, Cha GS, Ahn KH. Molecular recognition of anions through hydrogen bonding stabilization of anion–ionophore adducts: A novel trifluoroacetophenone-based binding motif. Organic Letters. 2003;**5**(21):4003-4006

[35] Chung YM, Raman B, Kim DS, Ahn KH. Fluorescence modulation in anion sensing by introducing intramolecular H-bonding interactions in host–guest adducts. Chemical Communications. 2006;(2):186-188

[36] Ryu D, Park E, Kim DS, Yan S, Lee JY, Chang BY, Ahn KH. A rational approach to fluorescence "turn-on" sensing of α-amino-carboxylates. Journal of the American Chemical Society. 2008;**130**(8):2394-2395

[37] Kim DS, Chung YM, Jun M, Ahn KH. Selective colorimetric sensing of anions in aqueous media through reversible covalent bonding. The Journal of Organic Chemistry. 2009;**74**(13):4849-4854

[38] Kim DS, Ahn KH. Fluorescence "turn-on" sensing of carboxylate anions with oligothiophene-based o-(carboxamido) trifluoroacetophenones. The Journal of Organic Chemistry. 2008;**73**(17):6831-6834

[39] Niu HT, Jiang X, He J, Cheng JP. A highly selective and synthetically facile aqueous-phase cyanide probe. Tetrahedron Letters. 2008:**49**(46);6521-6524

[40] Niu HT, Su D, Jiang X, Yang W, Yin Z, He J, Cheng JP. A simple yet highly selective colorimetric sensor for cyanide anion in an aqueous environment. Organic & Biomolecular Chemistry. 2008;**6**(17):3038-3040

[41] Ekmekci Z, Yilmaz MD, Akkaya EU. A monostyryl-boradiazaindacene (BODIPY) derivative as colorimetric and fluorescent probe for cyanide ions. Organic Letters. 2008;**10**(3):461-464

[42] Yu H, Zhao Q, Jiang Z, Qin J, Li Z. A ratiometric fluorescent probe for cyanide: convenient synthesis and the proposed mechanism. Sensors and Actuators B: Chemical. 2010;**148**(1):110-116

[43] Peng L, Wang M, Zhang G, Zhang D, Zhu D. A fluorescence turn-on detection of cyanide in aqueous solution based on the aggregation-induced emission. Organic Letters. 2009;**11**(9):1943-1946

[44] Sessler JL, Cho D. The benzil rearrangement reaction: Trapping of a hitherto minor product and its application to the development of a selective cyanide anion indicator. Organic Letters. 2008;**10**(1):73-75. DOI: 10.1021/ol7027306

[45] Miyaji H, Kim DS, Chang BY, Park E, Park SM, Ahn KH. Highly cooperative ion-pair recognition of potassium cyanide using a heteroditopic ferrocene-based crown ether–trifluoroacetylcarboxanilide receptor. Chemical Communications. 2008;(6):753-755

[46] Sun Y, Liu Y, Chen M, Guo W. A novel fluorescent and chromogenic probe for cyanide detection in water based on the nucleophilic addition of cyanide to imine group. Talanta. 2009;**80**(2):996-1000

[47] Sun Y, Liu Y, Guo W. Fluorescent and chromogenic probes bearing salicylaldehyde hydrazone functionality for cyanide detection in aqueous solution. Sensors and Actuators, B: Chemical. 2009;**143**(1):171-176

[48] Cho D, Kim JH, Sessler JL. The benzil–cyanide reaction and its application to the development of a selective cyanide anion indicator. American Chemical Society. 2008;**130**(36):12163-12167. DOI: 10.1021/ja8039025

[49] Kim HJ, Ko KC, Lee JH, Lee JY, Kim JS. KCN sensor: Unique chromogenic and 'turn-on' fluorescent chemodosimeter: Rapid response and high selectivity. Chemical Communications. 2011;**47**(10):2886-2888 (b) Lee JH, Jeong AR, Shin S, Kim HJ, Hong JI. Fluorescence turn-on sensor for cyanide based on a cobalt (II)– coumarinylsalen complex. Organic Letters. 2010;**12**(4):764-767 (c) Saha S, Ghosh A, Mahato P, Mishra S, Mishra SK, Suresh E, Das S, Das A. Specific recognition and sensing of CN– in sodium cyanide solution. Organic Letters. 2010;**12**(15):3406-3409 (d) Chen X, Zhou Y, Peng X, Yoon J. Fluorescent and colorimetric probes for detection of thiols. Chemical Society Reviews. 2010;**39**(6):2120-2135

[50] (a) Lee KS, Kim HJ, Kim GH, Shin I, Hong JI. Fluorescent chemodosimeter for selective detection of cyanide in water. Organic Letters. 2008;**10**(1):49-51 (b) Kumara V, Kaushika MP, Srivastavaa AK, Pratapa A, Thiruvenkatamb V, Rowb TG. Thiourea based novel chromogenic sensor for selective detection of fluoride and cyanide anions in organic and aqueous media. Analytica Chimica Acta. 2010;**663**(1):77-84 (c) Li H, Li B, Jin LY, Kan Y, Yin B. A rapid responsive and highly selective probe for cyanide in the aqueous environment. Tetrahedron. 2011;**67**(38):7348-7353 (d) Martínez-Máñez R, Sancenón F. Chemodosimeters and 3D inorganic functionalised hosts for the fluoro-chromogenic sensing of anions. Coordination Chemistry Reviews. 2006;**250**(23-24):3081-3093. DOI: 10.1016/j.ccr.2006.04.016 (e) Chung SY, Nam SW, Lim J, Park S, Yoon J. A highly selective cyanide sensing in water via fluorescence change and its application to in vivo imaging. Chemical Communications. 2009;(20):2866-2868 (f) Shang L, Jin L, Dong S. Sensitive turn-on fluorescent detection of cyanide based on the dissolution of fluorophore functionalized gold nanoparticles. Chemical Communications. 2009;(21):3077-3079 (g) Zeng Q, Cai P, Li Z, Qin J, Tang BZ. An imidazole-functionalized polyacetylene: Convenient synthesis and selective chemosensor for metal ions and cyanide. Chemical communications. 2008;(9):1094-1096 (h) Zhang P, Shi BB, Wei TB, Zhang YM, Lin Q, Yao H, You XM. A naphtholic Schiff base for highly selective sensing of cyanide via different channels in aqueous solution. Dyes and Pigments. 2013;**99**(3):857-862

[51] Federation WE. American Public Health Association. Standard methods for the examination of water and wastewater. Washington, DC, USA: American Public Health Association (APHA); 2005

[52] United States Environmental Protection Agency (EPA), Titrimetric and manual spectro-photometric determinative methods for cyanide, Method 9014, 1996, http://www.epa.gov/waste/hazard/testmethods/sw846/pdfs/9014.pdf [Accessed January 9, 2010]

[53] United States Environmental Protection Agency (EPA), Potentiometric determination of cyanide in aqueous samples and distillates with ion-selective electrode, Method 9213. 1996. www.epa.gov/waste/hazard/testmethods/sw846/pdfs/9213.pdf [Accessed January 9, 2010]

[54] United States Environmental Protection Agency (EPA), Available cyanide by flow injection, ligand exchange and amperometry, Method OIA-1677. 2004. http://www.epa.gov/waterscience/methods/method/cyanide/1677final.pdf [Accessed January 9, 2010]

[55] World Health Organization. Concise international chemical assessment document 61,

Hydrogen cyanide and cyanides: Human health aspects, Geneva. 2004. pp. 4-5. http://www.who.int/ipcs/publications/cicad/en/cicad61.pdf [accessed January 09, 2009]

[56] The Agency for Toxic Substances and Disease Registry, Toxicological profile for cyanide, Atlanta, GA, US Department of Health and Human Services. Analytical Methods for Determining Cyanide in Biological Samples. pp. 201-219. http://www.atsdr.cdc.gov/toxprofiles/tp8-c7.pdf [Accessed January 9, 2010]

[57] Mudder TI, Botz M, Smith A. Chemistry and treatment of cyanidation wastes. London, UK: Mining Journal Books. pp. 21-46

[58] Mak KK, Yanase H, Renneberg R. Cyanide fishing and cyanide detection in coral reef fish using chemical tests and biosensors. Biosensors and Bioelectronics. 2005;**20**(12):2581-2593. DOI: 10.1016/j.bios.2004.09.015

[59] Singh HB, Wasi N, Mehra MC. Detection and Determination of Cyanide—A Review. International Journal of Environmental Analytical Chemistry. 1986;**26**(2):115-136

[60] Out EO, Byerley JJ, Robinson CW. Ion chromatography of cyanide and metal cyanide complexes: A review. International Journal of Environmental Analytical Chemistry. 1996;**63**(1):81-90. DOI: 10.1080/03067319608039812

[61] Lindsay AE, Greenbaum AR, O'Hare D. Analytical techniques for cyanide in blood and published blood cyanide concentrations from healthy subjects and fire victims. Analytica Chimica Acta. 2004;**511**(2):185-195. DOI: 10.1016/j.aca.2004.02.006

[62] Xu Z, Chen X, Kim HN, Yoon J. Sensors for the optical detection of cyanide ion. Chemical Society Reviews. 2010;**39**(1):127-137

[63] Zelder FH, Männel-Croisé C. Recent Advances in the colorimetric detection of cyanide. CHIMIA International Journal for Chemistry. 2009;**63**(1):58-62. DOI: 10.2533/chimia.2009.58

[64] Luo J, Xie Z, Lam JW, Cheng L, Chen H, Qiu C, Kwok HS, Zhan XW, Liu YQ, Zhu DB, Tang BZ. Aggregation-induced emission of 1-methyl-1, 2, 3, 4, 5-pentaphenylsilole. Chemical Communications. 2001;(18):1740-1741

[65] Sun Y, Li Y, Ma X, Duan L. A turn-on fluorescent probe for cyanide based on aggregationof terthienyl and its application for bioimaging. Sensors and Actuators B: Chemical. 2016;**224**:648-653. DOI: 10.1016/j.snb.2015.10.057

[66] Chen X, Wang L, Yang X, Tang L, Zhou Y, Liu R, Qu J. A new aggregation-induced emission active fluorescent probe for sensitive detection of cyanide. Sensors and Actuators B. 2017;**241**:1043-1049

[67] Kim G, Kim H. Doubly activated coumarin as a colorimetric and fluorescent chemodosimeter for cyanide. Tetrahedron Letters. 2010;**51**(1):185-187. DOI: 10.1016/j.tetlet.2009.10.113

[68] Hu J-W, Lin W-C, Hsiao S-Y, Wu Y-H, Chen H-W, Chen K-Y. An indanedione-based

chemodosimeter for selective naked-eye and fluorogenic detection of cyanide. Sensors and Actuators B. 2016;**233**:510-519

[69]  Huo F, Zhang Y, Yue Y, Chao J, Zhang Y, Yin C. Isophorone-based aldehyde for "ratio-metric" detection of cyanide by hampering ESIPT. Dyes and Pigments. 2017;**143**:270-275

[70]  Goswami S, Manna A, Paul S, Das AK, Aich K, Nandi PK. Resonance-assisted hydrogen bonding induced nucleophilic addition to hamper ESIPT: Ratiometric detection of cyanide in aqueous media. Chemical Communications. 2013;**49**:2912

[71]  Wang L, Jiqing D, Cao D. A colorimetric and fluorescent probe containing diketopyrrolopyrrole and 1,3-indanedione for cyanide detection based on exciplex signaling mechanism. Sensors and Actuators B. 2014;**198**:455-461

[72]  Shahid M, Razi SS, Srivastava P, Ali R, Maiti B, Misra A. A useful scaffold based on acenaphthene exhibiting $Cu^{2+}$ induced excimer fluorescence and sensing cyanide via $Cu^{2+}$ displacement approach. Tetrahedron. 2012;**68**:9076-9084

[73]  Goswami S, Paul S, Manna A. FRET based selective and ratiometric 'naked-eye' detection of $CN^-$ in aqueous solution on fluorescein–Zn–naphthalene ensemble platform. Tetrahedron Letters. 2014;**55**:3946-3949

[74]  Chung S-Y, Nam S-W, Lim J, Park S, Yoon J. A highly selective cyanide sensing in water via fluorescence change and its application to in vivo imaging. Chemical Communications. 2009:2866-2868

[75]  Yanar U, Babür B, Pekyılmaz D, Yahaya I, Aydıner B, Dede Y, Seferoğlu Z. A fluorescent coumarin-thiophene hybrid as a ratiometric chemosensor for anions: Synthesis, photophysics, anion sensing and orbital interactions. Journal of Molecular Structure. 2016;**1108**:269-277

[76]  Saha S, Ghosh A, Mahato P, Mishra S, Mishra SK, Suresh E, Das S, Das A. Specific recognition and sensing of $CN^-$ in sodium cyanide solution. Organic Letters. 2010;**12**(15)

[77]  Hao Y, Nguyen KH, Zhang Y, Zhang G, Fan S, Li F, Guo C, Yuanyuan L, Song X, Peng Q, Liu Y-N, Maotian X. A highly selective and ratiometric fluorescent probe for cyanide byrationally altering the susceptible H-atom. Talanta. 2018;**176**:234-241

[78]  Mashraqui SH, Betkar R, Chandiramani M, Estarellas C, Frontera A. Design of a dual sensing highly selective cyanide chemodosimeter based on pyridinium ring chemistry. New Journal of Chemistry. 2011;**35**:57-60

[79]  Kwona SK, Kou S, Ha NK, Chen X, Hwang H, Nama S-W, So HK, Swamy KMK, Park S, Yoon J. Sensing cyanide ion via fluorescent change and its application to the microfluidic system. Tetrahedron Letters. 2008;**49**:4102-4105

[80]  Li J, Qi X, Wei W, Zuo G, Dong W. A red-emitting fluorescent and colorimetric dual channel sensor forcyanide based on a hybrid naphthopyran-benzothiazol in aqueous solution. Sensors and Actuators B. 2016;**232**:666-672

# 5

# Catalytic Degradation of Organic Dyes in Aqueous Medium

Muhammad Saeed, Muhammad Usman and
Atta ul Haq

## Abstract

Water pollution by the textile industry is an emerging issue. Textile industry is the major industrial sector which contributes to water pollution. Textile industry releases a huge amount of unfixed dyes in wastewater effluents. About 20% of the dye production all over the world is discharged as waste in industrial effluents by textile industry. These dyes are highly stable and colored substances which disturb the aqueous ecosystem significantly. Therefore, there is a need for methods to remove organic dyes from textile industrial effluents. Photo catalysis and catalytic wet oxidation are best practices for degradation of dyes in wastewater. In photo catalysis, the dye molecules can be completely degraded into inorganic non-toxic compounds by irradiation of the dye solution under visible or ultra-violet light in the presence of semiconductor metal-oxide photo catalysts. In catalytic wet oxidation, various metal-based catalysts in supported or unsupported form can be used as heterogeneous catalysts for degradation of dyes in the presence of oxygen or hydrogen peroxide. These processes have several preferences like easy separation of the catalyst from reaction mixture and recycling of the catalyst.

**Keywords:** catalysis, degradation, dyes, Langmuir-Hinshelwood mechanism

## 1. Introduction

Water is an important resource in our society. In our planet, Earth, less than a 0.7% of the total of water is fresh water and only 0.01% is available to be used. Today, some of the most discussed issues around the world are sanitation, soil, air and water pollution [1]. Wastewater can be divided into four broad categories, according to its source, namely domestic, industrial, public service and system loss/leakage. Among these, industrial wastewaters occupy a 42.4% of the total volume and households a 36.4% of volume. Industrial effluents are the major sources of water contamination. These industrial effluents contain a wide variety of complex, biodegradable and non-biodegradable organic substances like dyes, pesticides and herbicides and dyes with different concentrations [2]. A large amount of water is being used in the textile industry for preparing fabric and dyeing process. Other industries like plastics, pharmaceutical, pulp, leather and food industries also use dyes; however, textile industry is considered as a major water consumer as well as the largest dye consumer sector. The wastewater coming out of these industries contains a huge amount of dyes. Textile industry releases about 20% of the total world production of dyes in wastewater effluents.

Dyes are complex organic compounds which are used to impart color to materials. Dyes are categorized into different classes like anthraquinone, azo, reactive, disperse, acidic, basic and neutral dyes. Most commonly employed dyes are anthraquinone and azo dyes and more than 60% of these dyes are reactive dyes [3]. Reactive azo dyes are extensively used in the textile industry. Azo dyes consisting of a diazotized amine coupled to an amine or a phenol and contain one or more azo ($-N = N-$) linkage comprise about half of all textile dyes used in the present textile industry. The concentrations of these dyes in textile waste-water are significantly high, as about 20% of dye residues are released into effluents [4]. These industrial effluents are released directly into receiving waters without much effective treatment. The dye discharge into the environment poses serious threat to sustainability of ecosystems because they are highly resistant to natural degradation. These dyes have been declared as carcinogenic and tumorigenic material by the International Agency for Research on Cancer (IARC) and National Institute for Occupational Safety and Health; however, these dyes are still in use in textile dyeing processes. These dyes possess a complex nature due to a large degree of aromatics and synthetic origin. These dyes are highly stable, resistant towards photo and biological degradation and refractory against chemical oxidation. These characteristics make conventional biochemical and physiochemical techniques ineffective. Therefore, elimination of such dyes from textile effluents is of considerable interest and importance. Hence the textile effluents should be effectively treated before final disposal [5–7].

## 2. Photo catalytic degradation

Increase of recalcitrant organic pollutants in industrial effluents developed the law and regulations related to environment more forceful. As a response, advancement in new, more powerful and eco-friendly protocols for degradation of organic pollutants in industrial effluents turned into an important task. For the treatment of industrial effluents, various physical and chemical processes, such as ion exchange, adsorption, flocculation, UV radiation, electrochemical reduction, ozonisation and so on, have been used for elimination of dyes in the past few decades [8–12]. However, most of these processes face secondary pollution problems, complicated procedures and high cost. Hence, the development of an effective and eco-friendly protocol for treatment of industrial effluents is needed.

Ultimately, researchers focused on advanced oxidation processes (AOPs) to eliminate these stable pollutants from the aqueous medium. Advanced oxidation processes involve the production of active radicals like OH. These active radicals take part in decomposition of macromolecules of pollutants into less harmful and smaller substances [13–14]. Fenton process, sonolysis, ozonation process, radiation-induced degradation, biodegradation and heterogeneous photo catalysis are techniques which are employed in the AOP approach [15–19]. The heterogeneous photo catalysis, which can be used for degradation of organic pollutants by initiation of redox transformations, has been proved as an efficient tool for degradation of aqueous as well as atmospheric organic pollutants. This technique involves the initiation of photo reactions in the presence of a semiconductor photo catalyst. A number of photosensitive semiconductors such as ZnO, $V_2O_5$, ZnS, CdS, $TiO_2$, ZnO, oxides of Mn and so on can be employed as photo catalysts for aqueous-phase photo degradation of organic pollutants due to their environmental-friendly benefits [20–25].

## 3. Mechanism of photo catalytic degradation

Irradiation excites an electron from valence band to conduction band of the semiconductor photo catalyst leaving behind a positive hole in valence band. Electron from the conduction band is taken up by oxygen adsorbed at the surface of photo catalyst and produces superoxide anion ($O^{2-}$). This super oxide anion further reacts with water and produces the OH radical. Similarly, the positive hole moves to the surface of catalyst and reacts with water and produces OH radical. These OH radicals take part in decomposition of the pollutants. This mechanism has been illustrated in **Figure 1**.

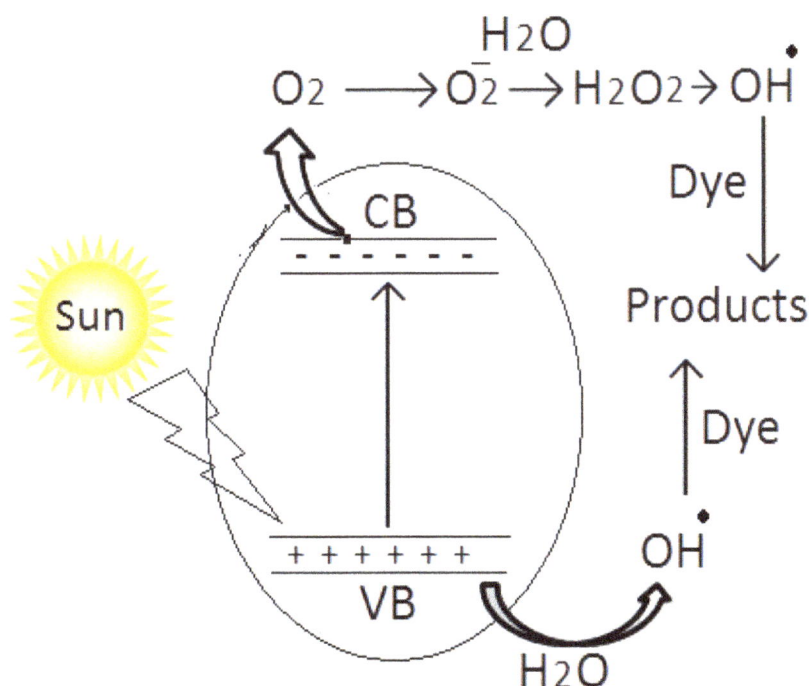

**Figure 1.** Mechanism of photo catalytic degradation of dyes.

The catalytic activities of photo catalysts towards degradation of dye molecules can be enhanced by doping of semiconductor metal-oxide photo catalysts with metal nanoparticles of low fermi level like Ag. These metal nanoparticles doped on semiconductor photo catalysts prevent the recombination of the electron–hole pair by the well-known Schottky barrier effect which results in an increase in quantum efficiency of the photo catalysts [26, 27]. It has been reported, for example, that doping of Ag greatly enhanced the photo catalytic activities of the manganese oxide photo catalyst for degradation of rhodamine B dye in aqueous medium as shown in **Figure 2** [28]. It was observed that 11% (in 15 min of reaction) and 45% (in 120 min of reaction) of 40 mL (200 ppm) solution of rhodamine B degraded when using manganese oxide as the photo catalyst at 40°C. In the presence of Ag-doped manganese oxide, 28% (in 15 min) and 91% (in 120 min) degradation of rhodamine B dye were achieved under similar reaction conditions. The pH of the solution also affects the catalytic activities of photo catalysts. Higher pH favors the production of OH radicals; therefore, the photo catalytic activities of photo catalysts increase with pH. **Figure 3** shows the effect of pH on photo catalytic activities of Ag-doped manganese oxide [28]. Similarly, Ansari et al. [29] have reported that $TiO_2$ and Ag-doped $TiO_2$ catalyzed photo degradation of methyl orange and methylene blue dyes. It was found that 42% and 88% degradation of methyl orange and methylene blue dye was achieved after 6 h of reaction. In the presence of Ag-doped $TiO_2$, the degradation achieved was 78 and 96% for methyl orange and methylene blue dyes.

## 3.1. Kinetics of photo catalytic degradation

Eley-Rideal (E-R) mechanism, one of the three mechanisms of heterogeneous catalytic reactions (Langmuir-Hinshelwood, Mars van Krevelen and Eley Rideal), can be used to describe the kinetics of the photo catalytic degradation of dyes in the presence of oxygen. Eley-Rideal (E-R) mechanism states that surface-catalyzed reaction proceeds in two steps. In the first step, the gaseous reactant, oxygen, gets adsorbed at the surface of the catalyst followed by a reaction

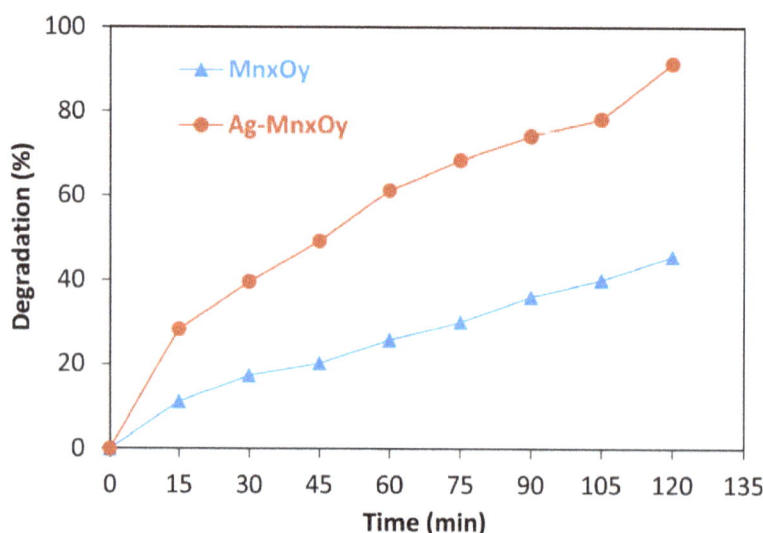

**Figure 2.** Effect of doping of Ag on photo catalytic activities of manganese oxide for photo degradation of rhodamine B dye in aqueous medium [28].

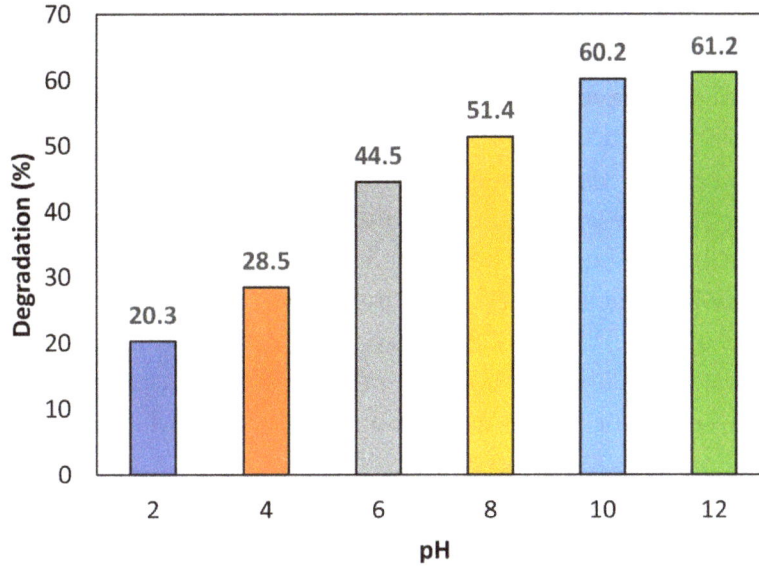

**Figure 3.** Effect of pH on photo catalytic activities of Ag-doped manganese oxide for photo degradation of rhodamine B dye in aqueous medium [28].

with a fluid phase reactant, the dye in second step. The adsorbed oxygen scavenges the electrons generated in the conduction band by the irradiation of catalyst and yields superoxide anions ($O^{2-}$). Super oxide anions transform to OH radicals by protonation. The positive holes in valence band also generate OH radicals by reaction with water. These OH radicals play a significant performance in the mineralization of B dye [29]. The rate expression for the Eley-Rideal mechanism can be written as:

$$-\frac{dR}{dt} = k_r R \theta_{O2} \tag{1}$$

R, $\theta_{O2}$ and $k_r$ indicate concentration of dye, surface concentration of oxygen and rate constant of reaction, respectively. Rate of reaction becomes independent of oxygen under constant flow of oxygen; hence

$$-\frac{dR}{dt} = k_{Ap} R \tag{2}$$

$k_{Ap}$ is the apparent rate constant,

Equation (2) can be expressed in integral form as

$$\ln \frac{R_o}{R_t} = k_{Ap}\, t \tag{3}$$

$R_0$ and $R_t$ is the concentration of dye at time zero and time t, respectively. The plot of $\ln(R_0/R_t)$ versus t gives a straight line. The slope of this straight line gives the apparent rate constant for photo catalytic degradation of dyes in the aqueous medium.

The time-profile data of Ag-doped manganese oxide catalyzed photo-degradation of rhodamine B dye and was subjected to kinetics analysis according to Eq. (3). It was noted that fitting Eq. (3) to experimental data gave the best straight lines as given in **Figure 4**. The apparent rate constants determined from the slopes of straight line were 0.0136, 0.0151 and 0.0216 per minute at 303, 313 and 323 K, respectively [28]. Similarly, we studied the ZnO-catalyzed photo degradation of methyl orange in aqueous medium [30]. The data obtained was subjected to kinetics analysis according to Eq. (3). The data gave a best fit to kinetics expression as given in **Figure 5**. The rate constants determined from the slopes of straight lines were 0.0098, 0.0128 and 0.0163 per minute at 303, 313 and 323 K, respectively. In another study [31], ZnO was used as a photo catalyst for degradation of rhodamine B dye in the aqueous medium. The experimental data was analysed according to expression 3. It was found that the data gave best fit to

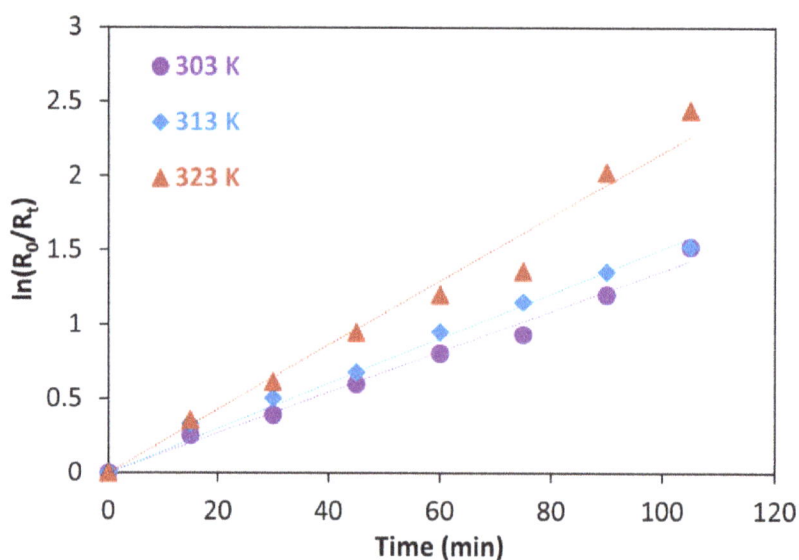

**Figure 4.** Kinetics of Ag-doped manganese oxide catalyzed photo degradation of rhodamine B dye in aqueous medium.

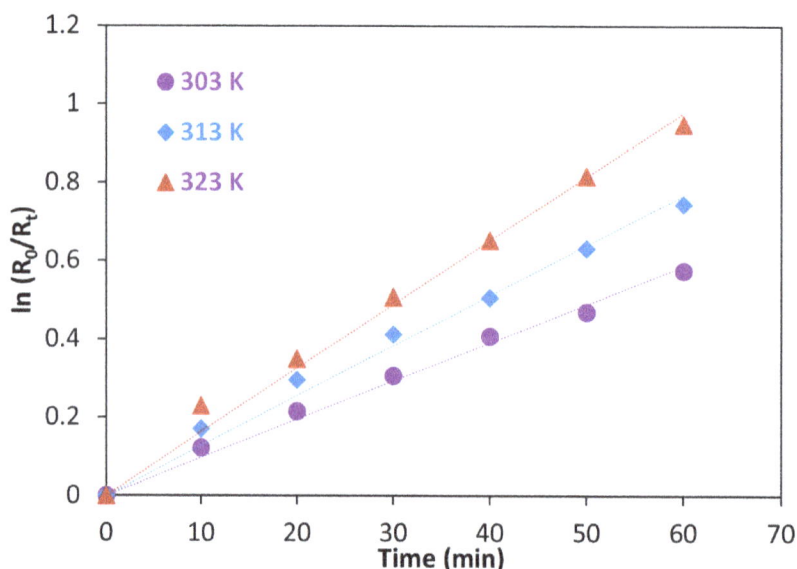

**Figure 5.** Kinetics of ZnO catalyzed photo degradation of methyl orange in aqueous medium.

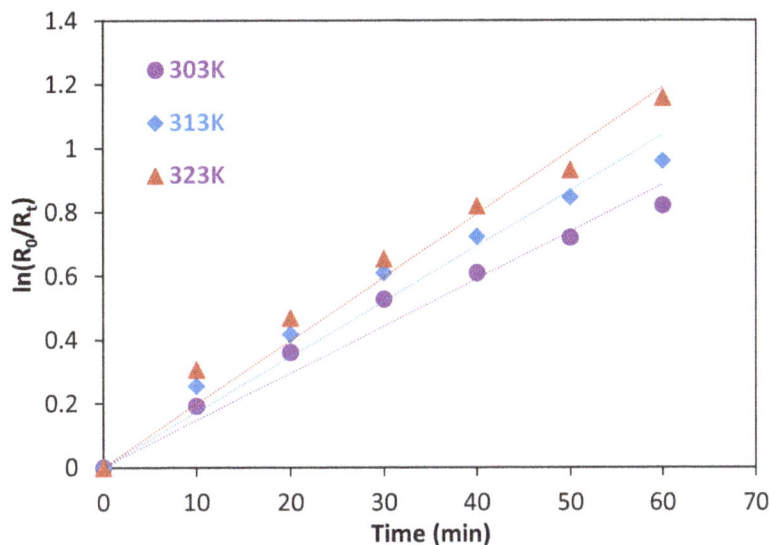

**Figure 6.** Kinetics of ZnO catalyzed photo degradation of rhodamine B dye in aqueous medium.

kinetic expression as given in **Figure 6**. The rate constants determined were 0.0148, 0.0174 and 0.0199 per minute at 303, 313 and 323 K, respectively.

# 4. Wet catalytic oxidation/degradation

In the recent years, the field of wastewater treatment and reuse of industrial processed water, for the suitable development of human activities, has achieved much attention. Chemical oxidation of organic wastes in water is one of the recommended and environmental-friendly methods to improve water quality. The wet-air or thermal liquid-phase oxidation process is known to have a great potential for the treatment of effluents containing organic toxic contaminants. The efficient removal of pollutants via the wet-air oxidation process requires very high temperature and pressure, typically in the range 473–573 K and 7–15 MPa, respectively, which leads to high installation costs, and practical applications for this process are limited. Therefore, the development of heterogeneous catalytic wet oxidation using various types of catalysts and oxidants has been attempted in order to reduce the severity of the oxidation conditions. The use of heterogeneous catalysts makes the process more attractive by achieving high efficiency for oxidation of organic wastes at considerably lower temperature and pressure. Compared to conventional wet oxidation, catalytic wet oxidation has lower energy requirements in which organic compounds are oxidized to inorganic compounds such as $CO_2$ and $H_2O$ [32–34].

The catalytic degradation of dyes is a slurry-phase reaction having reactants in liquid phase and catalysts in solid phase. It is generally assumed that surface-catalyzed reactions occur by a reaction between reactants adsorbed at the surface of the catalyst. The surface-catalyzed reaction can be broken down into the following basic five steps. Any one of these steps can be the rate determining step:

1.  Transport of reactants to the catalyst

2.  Adsorption of reactants at the catalyst surface

3.  Reaction between adsorbed reactants on the surface of catalyst

4.  Desorption of the products from the catalyst

5.  Transport of products away from the catalyst

Steps 2, 3 and 4 are chemical in nature. These steps jointly are regarded as the catalytic reaction. If any of these steps is slow step, that is, rate determining step, the reaction is said to be taking place in the kinetic-controlled region. Step 1 and 5, on the other hand, are physical processes which involve no chemical change. When either of these is slower, the reaction is said to be diffusion controlled or the rate is said to be diffusion limited. The diffusion control or kinetic control nature of the heterogeneous reaction can be confirmed by studying the effect of stirring on the rate of reaction [35, 36].

## 4.1. Kinetics of wet catalytic oxidation/degradation

As stated earlier, degradation of dyes takes place at the surface of catalyst. The kinetics of surface-catalyzed reaction can be described by one of the three possible mechanisms of heterogenous catalysis [37]:

1.  The Langmuir-Hinshelwood mechanism (L-H)

2.  The Mars-van Krevelen mechanism (M-K)

3.  The Eley-Rideal mechanism (E-R)

According to Langmuir-Hinshelwood (L-H), the reaction proceeds in two steps. In the first step the reactants get adsorbed on the surface of the catalyst and in the second step the adsorbed reactants react and give the final products. This whole process can be summarized in the following steps.

1.  $R + {}^* \rightarrow R^*$ Adsorption of dye on the surface of the catalyst

2.  $O_2 + {}^* \rightarrow O_2^*$ Adsorption of oxygen on catalyst surface

3.  $R^* + O_2^* \rightarrow P_s^*$ Reaction at the surface of catalyst

4.  $P_s^* \rightarrow P_s + {}^*$ Desorption of the products

According to the Langmuir-Hinshelwood (L-H) theory, the rate of reaction can be given by following equation.

$$Rate = k_r \theta_R \theta_{O_2} \tag{4}$$

where $\theta_R$ and $\theta_{O2}$ represent the surface covered by dye, R and molecular oxygen, respectively. Adsorption of dye, R and oxygen on the surface of the catalyst may take place according to

- Langmuir adsorption isotherm

- Temkin adsorption isotherm

- Freundlich adsorption isotherm

Langmuir adsorption isotherm may be either competitive or non-competitive. If adsorption of dye and oxygen takes place according to the competitive Langmuir adsorption isotherm, then fraction of the surface covered by reactant R and oxygen can be represented by expressions 5 and 6, respectively.

$$\theta_R = \frac{K_R[R]}{1 + K_R[R] + K_{O_2}[O_2]_g^n} \tag{5}$$

$$\theta_{O_2} = \frac{K_{O_2}[O_2]_g^n}{1 + K_R[R] + K_{O_2}[O_2]_g^n} \tag{6}$$

$K_R$ and $K_{O2}$ represent adsorption coefficient for dye R and oxygen, respectively. The value of n can be taken as 1 or 0.5 for molecular or dissociative adsorption of oxygen, respectively.

Putting the values of $\theta_R$ and $\theta_{O2}$ from Eqs. (5) and (6) in Eq. (4), we get Eq. (7).

$$Rate = k_r \frac{K_R[R]K_{O_2}[O_2]_g^n}{\left(1 + K_R[R] + K_{O_2}[O_2]_g^n\right)^2} \tag{7}$$

At constant flow of oxygen, the expression 7 transforms to.

$$Rate = \frac{ab[R]}{(c + b[R])^2} \tag{8}$$

where a, b and c are $k_r\,K_{O2}\,[O_2]$, $K_R$ and $1 + K_{O2}\,[O_2]$, respectively.

If the adsorption of reactant R and oxygen at the catalyst surface is taking place according to the non-competitive Langmuir adsorption isotherm, then fraction of the surface covered by reactant R and oxygen may be represented by expression 9 and 10, respectively.

$$\theta_R = \frac{K_R[R]}{1 + K_R[R]} \tag{9}$$

$$\theta_{O_2} = \frac{K_{O_2}[O_2]_g^n}{1 + K_{O_2}[O_2]_g^n} \tag{10}$$

Putting the values of $\theta_R$ and $\theta_{O2}$ from Eqs. (9) and (10) in Eq. (4), we get.

$$Rate = k_r \frac{K_R[R]K_{O_2}[O_2]_g^n}{(1 + K_R[R])(1 + K_{O_2}[O_2])} \tag{11}$$

At constant flow/pressure of oxygen, the expression 11 transforms to Eq. (12).

$$Rate = \frac{ab[R]}{1 + b[R]} \tag{12}$$

Similarly, if adsorption of dye and oxygen at the surface of catalyst follow Temkin or Freundlich isotherm, then the rate expression becomes Eqs. (13) and (14), respectively.

$$Rate = \bar{k}_r(K_1 \ln K_2[R]) \tag{13}$$

$$Rate = \bar{k}_r K_R[R]^{1/n} \tag{14}$$

Like Langmuir-Hinshelwood mechanism, Mars-van Krevelen mechanism also comprises of two steps. In the first step, the lattice oxygen of the catalyst oxidizes the substrate molecule and hence produces a partially reduced catalyst. In the second step, the reduced catalyst is reoxidised by molecular oxygen. The rate equation for Mars-van Krevelen can be given by expression 15.

$$Rate = \frac{k_1[R]k_2[O_2]_g^n}{\beta k_1[R] + k_2[O_2]_g^n} \tag{15}$$

$k_1$, $k_2$ and $\beta$ is the rate constant for degradation of dye, R is the rate constant for reoxidation of catalyst and stoichiometric coefficient of oxygen (0.5), respectively. At constant flow of oxygen, expression 15 changes to expression 16.

$$Rate = \frac{a[R]}{b + c[R]} \tag{16}$$

According to the Eley-Rideal mechanism (E-R) mechanism, the gaseous reactant gets adsorbed on the surface of the catalyst while the second reactant, dye, reacts with the adsorbed reactant from the fluid phase. In the present case, oxygen is adsorbed at the surface while reactant R remains in the fluid phase.

Rate expression for the Eley-Rideal mechanism (E-R) can be given by Eq. (17) as below:

$$Rate = k_r \theta_{O_2}[R] \tag{17}$$

In case of constant pressure of oxygen, the above equation can be transformed to Eq. (18) by lumping all the constants together as given below.

$$Rate = a[R] \tag{18}$$

All these equations can be applied to experimental data of heterogenous catalytic degradation of dyes in the aqueous medium by linear and non-linear method of analysis. We studied the degradation of rhodamine B and methylene blue dyes catalyzed by $CoFe_2O_4$ in aqueous medium [38]. Th data obtained was subjected to kinetic analysis according to above equations using Curve Expert software. It was found that Eq. (12) was best applicable to the data indicating that reactions were taking place according to the Langmuir-Hinshelwood mechanism. The constants determined using Curve Expert are given in **Table 1**. In another study [39] we employed nickel hydroxide as catalyst for degradation of black dye in aqueous medium. The experimental data was analysed according to kinetics discussed above. It was found that the reaction followed the Langmuir-Hinshelwood mechanism. The constants determined by Curve Expert software are listed in **Table 2**. Similarly, the degradation of methylene blue dye catalyzed by nickel hydroxide was also investigated [40]. The experimental data was analysed using the Curve Expert software. The Langmuir-Hinshelwood mechanism was applicable in this study as well. The rate constants and adsorption equilibrium constants are listed in **Table 3**.

| T (K) | $k_r$ (/min) | | $k_x$ (L/mol) | |
|---|---|---|---|---|
| | RhB | MB | RhB | MB |
| 303 | 0.028 | 0.031 | 1.61 | 1.74 |
| 313 | 0.066 | 0.069 | 1.60 | 1.71 |
| 323 | 0.098 | 0.098 | 1.58 | 1.69 |
| 333 | 0.128 | 0.131 | 1.55 | 1.67 |

**Table 1.** Kinetics parameter determined by application of Langmuir model (Eq. 12) to time profile data of $CoFe_2O_4$ catalyzed degradation of rhodamine B and methylene blue dyes dyes using Curve Expert software.

| T (K) | $k_r$ (/min) | $k_x$ (L/mol) |
|---|---|---|
| 313 | 0.051 | 0.153 |
| 323 | 0.093 | 0.125 |
| 333 | 0.218 | 0.079 |

**Table 2.** Kinetics parameter determined by application of Langmuir model (Eq. 12) to time profile data of nickel hydroxide catalyzed degradation of direct black dye using Curve Expert software.

| T (K) | $k_r$ (/min) | $k_x$ (L/mol) |
|---|---|---|
| 313 | 1.12 | 0.0112 |
| 323 | 1.55 | 0.0093 |
| 333 | 2.05 | 0.0084 |

**Table 3.** Kinetics parameter determined by application of Langmuir model (Eq. 12) to time profile data of nickel hydroxide catalyzed degradation of methylene blue dye using Curve Expert software.

# 5. Conclusion

Water pollution by textile industry is an emerging issue. Textile industry releases about 20% of the total world production of dyes in wastewater effluents. Photo catalysis and catalytic wet oxidation are best techniques for elimination of these dyes from aqueous system. In photo catalysis, the dye molecules are degraded to inorganic non-toxic compounds by irradiation of dye solution under visible or ultraviolet light in the presence of semiconductor metal-oxide photo catalysts. In catalytic wet oxidation, various metal-based catalysts in supported or unsupported form can be used as heterogeneous catalysts for degradation of dyes in the presence of oxygen or hydrogen peroxide. These processes have several preferences like easy separation of catalysts from the reaction mixture and recycling of the catalyst.

# Author details

Muhammad Saeed*, Muhammad Usman and Atta ul Haq

*Address all correspondence to: msaeed@gcuf.edu.pk

Department of Chemistry, Government College University Faisalabad, Faisalabad, Pakistan

# References

[1] Maldonado MC, Campo T, Elizalde E, Morant C, Marquez F. Photocatalytic degradation of rhodamine-B under UV-visible light irradiation using different nanostructured catalysts. American Chemical Science Journal. 2013;**3**(3):654-661

[2] Alinsafi A, Evenou F, Abdulkarim M, Pons MN, Zahraa O, Benhammou A, Yaacoubi A, Nejmeddine A. Treatment of textileindustry wastewater by supported photocatalysis. Dyes and Pigments. 2007;**74**:439-445

[3] Siddique M, Farooq R, Abda K, Athar F, Qaisar M, Umar F, Raja IA, Shaukat SF. Thermal-pressure mediated hydrolysis of reactive blue 19 dye. Journal of Hazardous Material. 2009;**172**:1007-1012

[4] Galindo C, Jacques P, Kalt A. Photo-oxidation of the phenylazonapthol AO20 on $TiO_2$: Kinetic and mechanistic investigations. Chemosphere. 2001;**45**(6–7):997-1005

[5] Byrappa K, Subramani AK, Ananda S, Rai KML, Dinesh R, Yoshimura M. Photocatalytic degradation of rhodamine B dye using hydrothermally synthesized ZnO. Bulletin Material of Science. 2006;**29**:433-438

[6] Cuiping B, Xianfeng X, Wenqi G, Dexin F, Mo X, Zhongxue G, Nian X. Removal of rhodamine B by ozone-based advanced oxidation process. Desalination. 2011;**278**:84-90

[7] Kim KH, Ihm SK. Heterogeneous catalytic wet air oxidation of refractory organic pollutants in industrial wastewaters: A review. Journal of Hazardous Materials. 2011;**186**:16-34

[8] Neppolian B, Choi HC, Sakthivel S, Arabindoo B, Murugesan V. Solar/UV-induced photocatalytic degradation of three commercial textile dyes. Journal of Hazardous Material B. 2002;**89**:303-317

[9] Ma Y, Ding Q, Yang L, Zhang L, Shen Y. Ag nanoparticles as multifunctional SERS substrate for the adsorption: Degradation and detection of dye molecules. Applied Surface Science. 2013;**265**:346-351

[10] Gibson LT. Mesosilica materials and organic pollutant adsorption: Part B removal from aqueous solution. Chemical Society Reviews. 2014;**43**:5173-5182

[11] Gupta VK, Jain R, Varshney S. Electrochemical removal of the hazardous dye reactofix red 3 BFN from industrial effluents. Journal of Colloids and Interface Science. 2007;**312**:292-296

[12] Lei X, Chen C, Li X, Xue X, Yang H. Study on ultrasonic degradation of methyl orange wastewater by modified steel slag. Applied Mechanics and Materials. 2014;**662**:125-128

[13] Gaya UI, Abdullah AH. Heterogeneous photocatalytic degradation of organic contaminants over titanium dioxide: A review of fundamentals, progress and problems. Journal of Photochemistry and Photobiology C: Photochemistry Review. 2008;**9**:1-12

[14] Rauf MA, Meetani MA, Hisaindee S. An overview on the photocatalytic degradation of azo dyes in the presence of $TiO_2$ doped with selective transition metals. Desalination. 2011;**276**(1–3):13-27

[15] Song S, Ying H, He Z, Chen J. Mechanism of decolorization and degradation of CI direct red 23 by ozonation combined with sonolysis. Chemosphere. 2007;**66**:1782-1788

[16] Tehrani-Bagha AR, Mahmoodi NM, Menger FM. Degradation of a persistent organic dye from colored textile wastewater by ozonation. Desalination. 2010;**260**:34-38

[17] Vahdat A, Bahrami SH, Arami M, Motahari A. Decomposition and decoloration of a direct dye by electron beam radiation. Radiation Physics and Chemistry. 2010;**79**:33-35

[18] Ayed L, Chaieb K, Cheref A, Bakhrouf A. Biodegradation and decolorization of triphenylmethane dyes by Staphylococcus epidermidis. Desalination. 2010;**260**:137-146

[19] Elmorsi TM, Riyad YM, Mohamed ZH, Abd El Bary HMH. Decolorization of mordant red 73 azo dye in water using H2O2/UV and photo-Fenton treatment. Journal of Hazardous Material. 2010;**174**:352-358

[20] Mohamed MM, Al-Esaimi MM. Characterization, adsorption and photocatalytic activity of vanadium-doped $TiO_2$ and sulfated $TiO_2$ (rutile) catalysts: Degradation of methylene blue dye. Journal of Molecular Catalysis A Chemical. 2006;**255**:53-61

[21] Chen C, Wang Z, Ruan S, Zou B, Zhao M, Wu F. Photocatalytic degradation of C.I. acid orange 52 in the presence of Zn-doped $TiO_2$ prepared by a stearic acid gel method. Dyes Pigment. 2008;**77**:204-209

[22] Fujishima A, Zhang X, Tryk DA. $TiO_2$ photocatalysis and related surface phenomena. Surface Science Reports. 2008;**63**:515-582

[23] Andronic L, Enesca A, Vladuta C, Duta A. Photocatalytic activity of cadmium doped $TiO_2$ films for photocatalytic degradation of dyes. Chemical Engineering Journal. 2009;**152**:64-71

[24] El-Bahy ZM, Ismail AA, Mohamed RM. Enhancement of titania by doping rare earth for photodegradation of organic dye (direct blue). Journal of Hazardous Materials. 2009;**166**: 138-143

[25] Luo S, Duan L, Sun B, Wei m LX, Xu A. Manganese oxide octahedral molecular sieve (OMS-2) as an effective catalyst for degradation of organic dyes in aqueous solutions in the presence of peroxymonosulfate. Applied Catalysis B Environmental. 2015;**164**:92-99

[26] Liu Z, Ma C, Cai Q, Hong T, Guo K, Yan L. Promising cobalt oxide and cobalt oxide/silver photocathodes for photoelectrochemical water splitting. Solar Energy Materials and Solar Cells. 2017;**161**:46-51

[27] Yang X, Xu LL, Yu XD, Guo YH. One-step preparation of silver and indiumoxide co-doped $TiO_2$ photocatalyst for the degradation of rhodamine B. Catalysis Communications. 2008;**9**:1224-1229

[28] Saeed M, Ahmad A, Boddula R, Din I, Haq A, Azhar A. Ag@MnxOy: An effective catalyst for photo-degradation of rhodamine B dye. Environmental Chemistry Letters. 2018;**16**: 287-294. DOI: 10.1007/s10311-017-0661-z

[29] Ansari SA, Khan MM, Ansari MO, Cho MH. Silver nanoparticles and defect-induced visible light photocatalytic and photoelectrochemical performance of Ag@m-$TiO_2$ nanocomposite. Solar Energy Materials & Solar Cells. 2015;**141**:162-170

[30] Saeed M, Adeel S, Raoof HA, Usman M, Mansha A, Ahmad A. ZnO catalyzed degradation of methyl orange in aqueous medium. Chiang Mai Journal of Science. 2017;**44**(4): 1646-1653

[31] Saeed M, Siddique M, Usman M, Haq A, Khan SG, Raouf A. Synthesis and characterization of zinc oxide and evaluation of its catalytic activities for oxidative degradation of rhodamine B dye in aqueous medium. Zeitschrift Fur Physikalische Chemie. 2017;**231**(9): 1559-1572

[32] Rahman QI, Ahmad M, Misra SK, Lohani M. Effective photocatalytic degradation of rhodamine B dye by ZnO nanoparticles. Materials Letter. 2013;**91**:170-175

[33] Fu H, Pan C, Yao W, Zhu Y. Visible-light-induced degradation of rhodamine B by nanosized $Bi_2WO_6$. Journal of Physical Chemistry B. 2005;**109**:22432-22439

[34] Bhargava SK, Tardio J, Prasad J, Folger K, Akolekar DB, Grocott SC. Wet oxidation and catalytic wet oxidation. Industrial Engineering & Chemical Research. 2006;**45**:1221-1258

[35] Saeed M, Ilyas M. Oxidative removal of phenol from water catalyzed by lab prepared nickel hydroxide. Applied Catalysis B Environmental. 2013;**129**:247-254

[36] Saeed M, lyas M, Siddique M, Ahmad A. Oxidative degradation of oxalic acid in aqueous medium using manganese oxide as catalyst at ambient temperature and pressure. Arabian Journal for Science and Engineering. 2013;**38**:1739-1748

[37] Saeed M, Ilyas M, Siddique M. Kinetics of lab prepared manganese oxide catalyzed oxidation of benzyl alcohol in the liquid phase. International Journal of Chemical Kinetics. 2015;**47**(7):447-460

[38] Saeed M, Mansha A, Hamayun M, Ahmad A, Haq A, Ashfaq M. Green synthesis of $CoFe_2O_4$ and investigation of its catalytic efficiency for degradation of dyes in aqueous medium. Zeitschrift Fur Physikalische Chemie. 2018;**232**:359-371. DOI: 10.1515/zpch-2017-1065

[39] Saeed M, Haq A, Muneer M, Adeel S, Hamayun M, Ismail M. Degradation of direct black 38 dye catalyzed by lab prepared nickel hydroxide in aqueous medium. Global NEST Journal. 2016;**18**(2):309-320

[40] Saeed M, Jamal MA, Haq A, Younas M, Shahzad MA. Oxidative degradation of methylene blue in aqueous medium catalyzed by lab prepared nickel hydroxide. International Journal of Chemical Reactor Engineering. 2016;**14**(1):45-51

# 6

# Tailoring the Photophysical Signatures of BODIPY Dyes: Toward Fluorescence Standards across the Visible Spectral Region

Rebeca Sola Llano, Edurne Avellanal Zaballa,
Jorge Bañuelos,
César Fernando Azael Gómez Durán,
José Luis Belmonte Vázquez,
Eduardo Peña Cabrera and Iñigo López Arbeloa

**Abstract**

The modern synthetic routes in organic chemistry, as well as the recent advances in high-resolution spectroscopic and microscopic techniques, have awakened a renewable interest in the development of organic fluorophores. In this regard, boron-dipyrrin (BODIPY) dyes are ranked at the top position as luminophores to be applied in photonics or biophotonics. This chromophore outstands not only by its excellent and tunable photophysical signatures, but also by the chemical versatility of its core, which is readily available to a myriad of functionalization routes. In this chapter, we show that, after a rational design, bright and photostable BODIPYs can be achieved along the whole visible spectral region, being suitable as molecular probes or active media of lasers. Alternatively, the selective functionalization of the dipyrrin core, mainly at *meso* position, can induce new photophysical phenomena (such as charge transfer) paving the way to the development of fluorescent sensors, where the fluorescent response is sensitive to a specific environmental property.

**Keywords:** fluorescent probes, laser dyes, BODIPY, organic synthesis, charge transfer, fluorescent sensors

## 1. Introduction

The recent advances in high-resolution fluorescence microscopy have played a key role in the success and boom of bioimaging techniques in the last few years. Bioimaging is a noninvasive method to visualize biological processes by means of fluorescence emission in real time. Nowadays, thanks to super resolution microscopy (nanoscopy, awarded with the Nobel Prize of 2014 in chemistry), the diffraction resolution limit is surpassed, allowing the detection at the single molecular level [1–3]. The development of this revolutionary technological tool has awakened a renewed interest in the design of organic luminophores with improved properties as fluorescent probes or markers to detect or monitor biochemical events [4]. These molecules should fulfill some requirements to fit the requests of these advanced techniques. They should be biocompatible and photostable, and outstand by a high light absorption and emission probability, to guarantee long-lasting and bright fluorescence images. Note that these characteristics are almost the same than those required for the active media of tunable dye lasers [5, 6]. In other words, those molecules suitable as fluorescent probes are also valid as laser dyes.

The modern avenues in organic synthesis allow the development of organic fluorophores fulfilling the above requirements as well as with the required functionality to recognize selectively the target biomolecule [7]. Actually, there is a wide pool of commercially available families of organic dyes with absorption/emission bands covering almost the whole ultraviolet/visible spectral region. Among them, boron-dipyrromethene (BODIPY) dyes are likely at the "pole" position, owing to their excellent photophysical signatures [8, 9], but mainly due to the chemical versatility of its boron-dipyrrin core, ready available to a multitude of synthetic routes (**Figure 1**) [10, 11]. As a consequence, the BODIPY dyes can be exhaustively and specifically functionalized with a plethora of substituents [12–14]. At the same time, such substitution pattern rules the photophysical properties of the resulting dye or can induce new photophysical phenomena, such as charge or electron transfer, energy transfer or triplet state population. In other words, after a rational molecular approach, tailor-made BODIPYs can be designed with the specific requirements of any application demanding an organic dye. In fact,

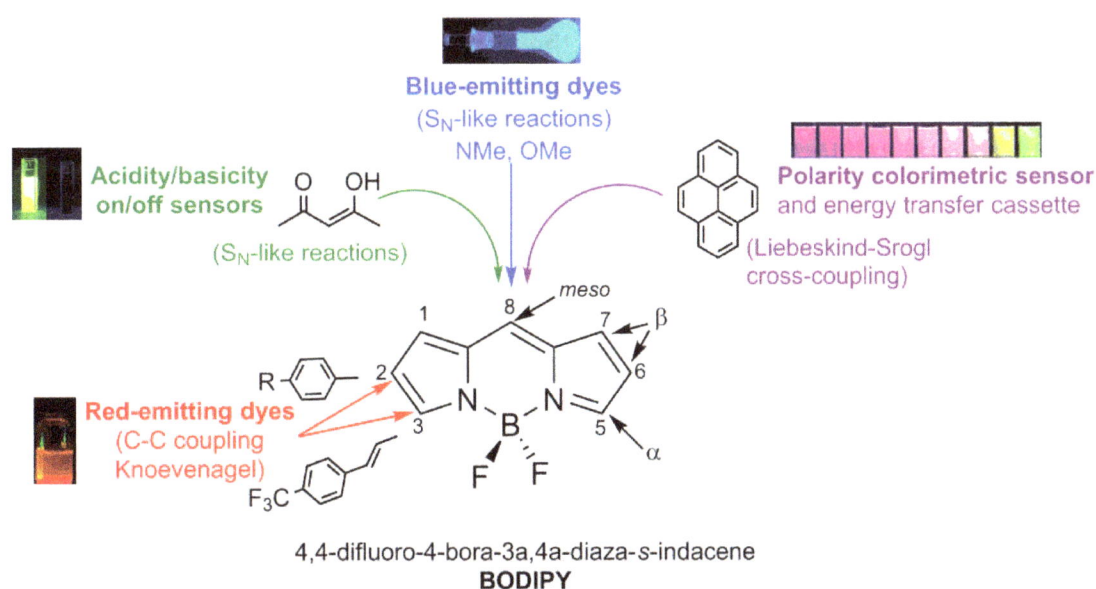

**Figure 1.** Basic molecular structure of the BODIPY dye. Schematic view of the structural modifications and chemical reactions tested herein.

a fast look at the bibliography reveals the popularity of these dyes since the number of publications dealing with this kind of dyes in many different applications fields is growing unstoppable during the last years. As a matter of fact, BODIPYs are not only applied in bioimaging [15] or as lasers [16], but also in many other scientific areas, such as photovoltaic devices as photosensitizers [17, 18], electrochemistry [19], sensors for recognition at the molecular level [20], or biomedicine in photodynamic therapy for the treatment of cancer [21, 22], just to cite some of the most relevant application fields.

In this chapter, we describe suitable molecular strategies to finely modulate the photophysical properties of BODIPY dyes (**Figure 1**). In the first part, we highlight some structural guidelines to shift the spectral bands to both edges (blue and red) of the visible spectral region, while keeping their characteristic high fluorescent efficiencies. In this way, fluorescent probes and laser dyes based on BODIPY can be attained covering the whole visible spectral region just choosing the adequate substitution pattern at the right chromophoric position. In the second part, we modify the molecular structure to induce new photophysical phenomena (energy and mainly charge transfer). To this aim, we selected the *meso* position since it is highly sensitive to the substituent effect. As a result, the fluorescent efficiency can response selectively to a specific property of the surrounding environment (such as polarity and acidity/basicity) leading to fluorescent sensors [23–25].

## 2. Modulation of the spectral shift

The structure of the BODIPY core features two pyrroles linked by a methine unit and a difluoroboron bridge (**Figure 2**). These dyes can be classified as polymethine dyes with a cyclic cyanine delocalized $\pi$-system leading to a quasi-aromatic backbone since the said boron center does not take part in the electronic delocalization [26]. Owing to symmetric reasons, the molecular dipole moment is oriented along the transversal axis and the negative charge is located around the fluorine atoms, whereas the positive charge around the opposite *meso* position (**Figure 2**). Thus, the chromophore core is fully planar and rigid. Such molecular structure ensures a high radiative deactivation probability, and a low nonradiative deactivation relaxation, since the internal conversion (related with the molecular flexibility), as well as the intersystem crossing probability (related with the spin-orbit coupling feasible in aromatic frameworks where the electron flow makes a loop), are largely truncated [27]. As a consequence, and taking as reference the simplest BODIPY with just hydrogens in the dipyrrin core [28], sharp and strong absorption and fluorescence bands are recorded in the green-yellow part (500–510 nm) of the visible (Vis) spectrum (HOMO → LUMO transition with the corresponding dipole moment oriented in the long molecular axis, **Figure 2**), with absorption coefficients approaching $10^5$ M$^{-1}$ cm$^{-1}$ and fluorescent efficiencies approaching the 100%, regardless of the solvent [9]. Besides, it shows stable laser emission at 540 nm with efficiencies approaching the 55% [29].

In the following sections, we show the tested strategies to induce deep hypsochromic and bathochromic shifts of the absorption and fluorescence spectral bands, while keeping high

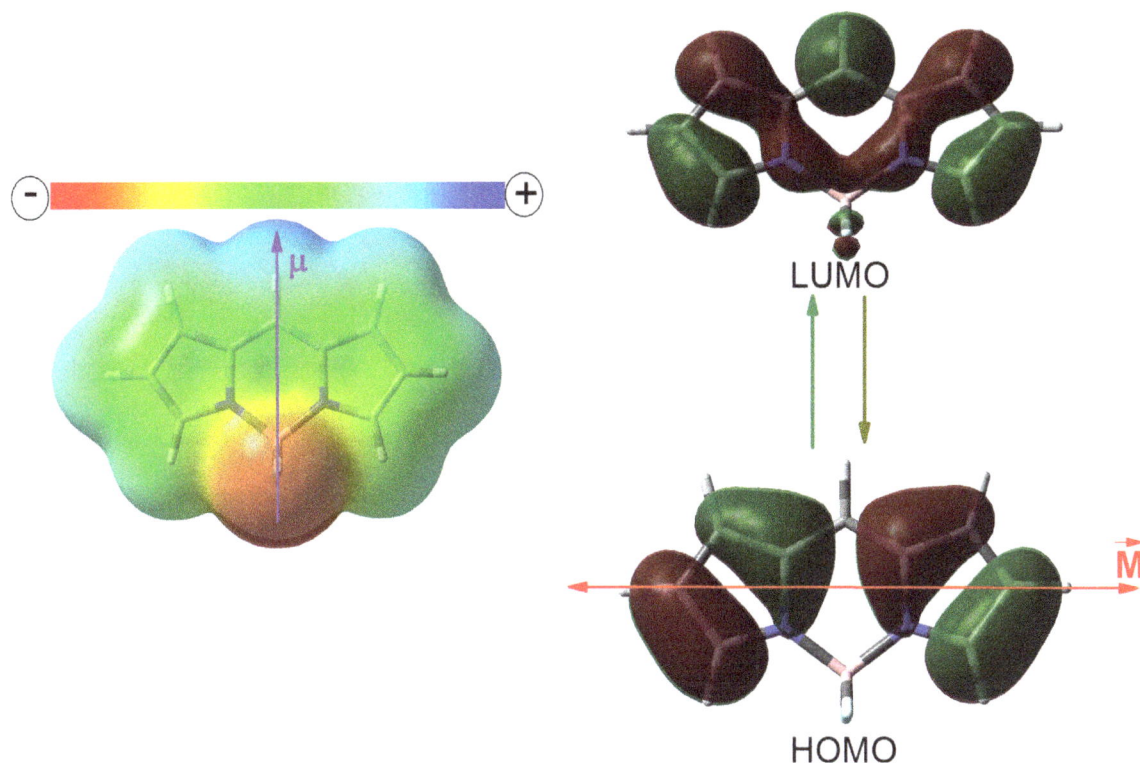

**Figure 2.** Frontier molecular orbitals and charge distribution by means of the electrostatic potential mapped onto the electronic distribution. The direction of the molecular dipole moment ($\mu$) as well as the transition dipole moment (M) is also depicted.

fluorescent efficiencies (see molecular structures in **Figure 3**). The main aim is to attain fluorophores based on BODIPY with optimal photophysical properties along the visible spectral region, ousting those commercially available as standard dyes, in terms on their fluorescence and laser performance, in each spectral region.

## 2.1. Blue-emitting BODIPYs

In the bibliography, some attempts have been made to achieve blue-emitting dyes based on BODIPY, but the core was drastically changed and the fluorescent response was not the best [30, 31]. However, strikingly, the sole incorporation of heteroatoms (nitrogen or oxygen, **Figure 3**) directly linked to the *meso* position led to the searched spectral hypsochromic shift (**Figure 4**) [29]. Such chromophoric position is highly sensitive to the substituent effect since a marked change in the electronic density takes place there upon excitation (note that in the HOMO, there is a node at such position, while it contributes much to the LUMO, **Figure 2**). This kind of 8-heteroatomBODIPYs can be straightforwardly attained after nucleophilic substitution at the 8-methylthioBODIPY precursor (provided by "cuantico" enterprise) [29]. Such starting precursor plays a key role in the chemistry of BODIPY since undergoes organic reactions unavailable until recently and open the door to new functionalization routes [32]. Thus, amines and alcohols can react specifically at *meso* position via nucleophilic substitution. The recorded spectral bands show that the hypsochromic shift depends on the electronegativity

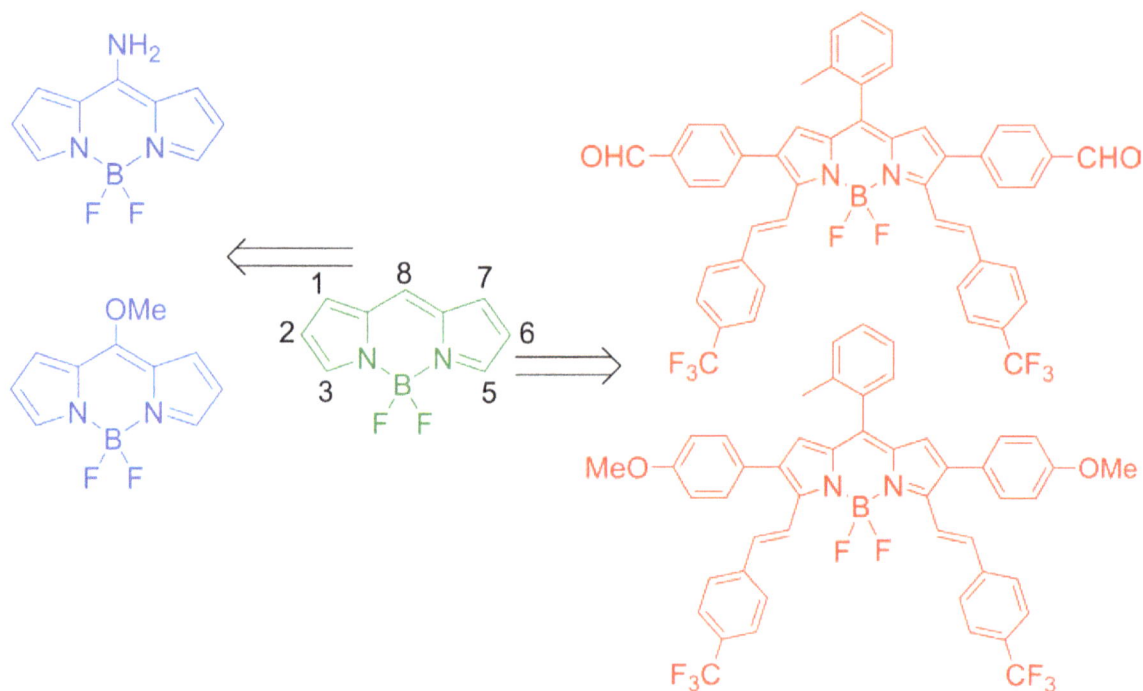

**Figure 3.** Molecular structure of the reference BODIPY dye and its blue-emitting (8-heteroatoms) and red-emitting (π-extended) derivatives.

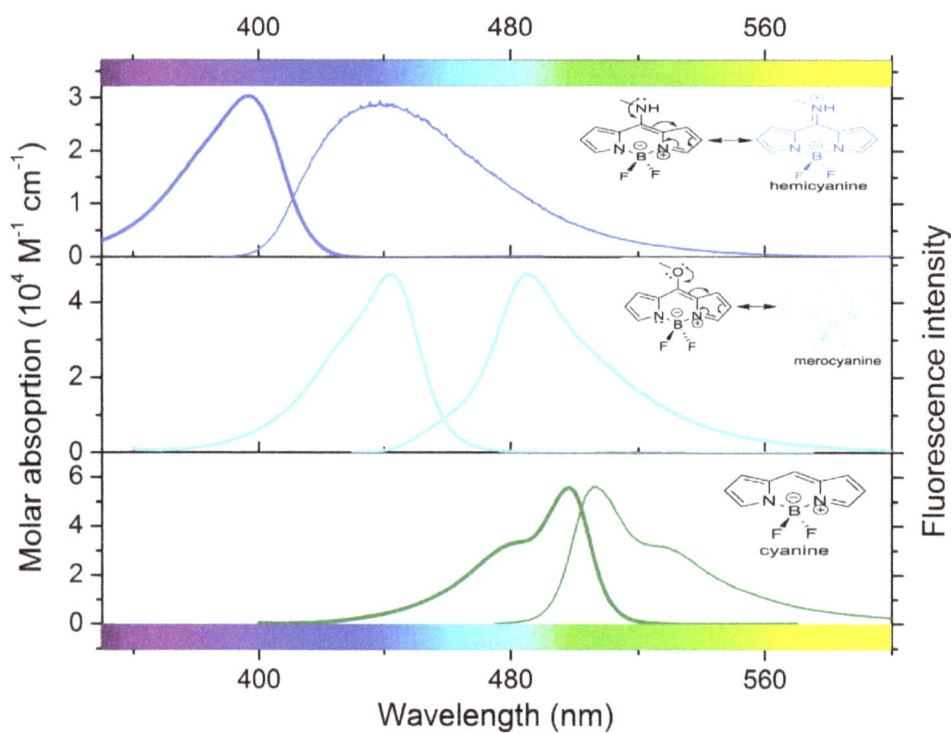

**Figure 4.** Absorption (bold line) and normalized fluorescence (thin line) spectra of the simplest BODIPY and its derivatives bearing methylamine and methoxy at *meso* position in a common solvent (ethanol). The main resonant structures resulting from the electronic coupling between the 8-heteroatoms and the dipyrrin core are also enclosed.

(or the electron releasing ability) of the attached heteroatom. Thus, while oxygen induces a moderate spectral shift (around 50 nm in absorption and 25 nm in fluorescence with regard to the reference simplest BODIPY), the nitrogen provokes a much more marked spectral shift (90 and 50 nm, respectively) placing the absorption/emission bands of the BODIPY in the blue-edge (420 and 460 nm, respectively) of the visible spectrum (**Figure 4**). Quantum mechanics calculations anticipate a hypsochromic shift upon the presence of heteroatoms, since their electron donor character raises the energy of the LUMO, increasing the energy gap. However, the theoretically predicted shift is much lower than the experimentally recorded, suggesting that additional processes are taking are taking place. Indeed, an electronic coupling between the heteroatom and one chromophoric pyrrole is feasible. Such resonant interaction seems to prevail mainly in the ground state and implies a reorganization of the electronic density leading to the stabilization of new delocalized π-systems, alternatively to the expected cyanine one; a hemicyanine and merocyanine, formed after electron coupling of the electron donor amine and methoxy, respectively (**Figure 4**). Thus, the hemicyanine is the responsible of the blue absorption, whereas the merocyanine of the greener one, owing to the lower electronegativity (more electron donor ability) of the nitrogen atom.

The BODIPYs bearing 8-methoxy are highly fluorescent regardless of the properties of the surrounding environment (**Figure 5**). In contrast, the fluorescent response of the 8-aminated analogs depends on the solvent polarity, being much lower in polar media. Such fluorescence quenching (from 70% in apolar to 15% in polar solvents, **Figure 5**) suggests that non-fluorescence intramolecular charge transfer (ICT) phenomena are switched on. The highly electron donor amine pushed the spectral bands toward the blue edge; however, such electron releasing ability can induce also the formation of an ICT state where the dipyrrin acts an electron acceptor. Note that the electron donor ability of the methoxy seems not high enough

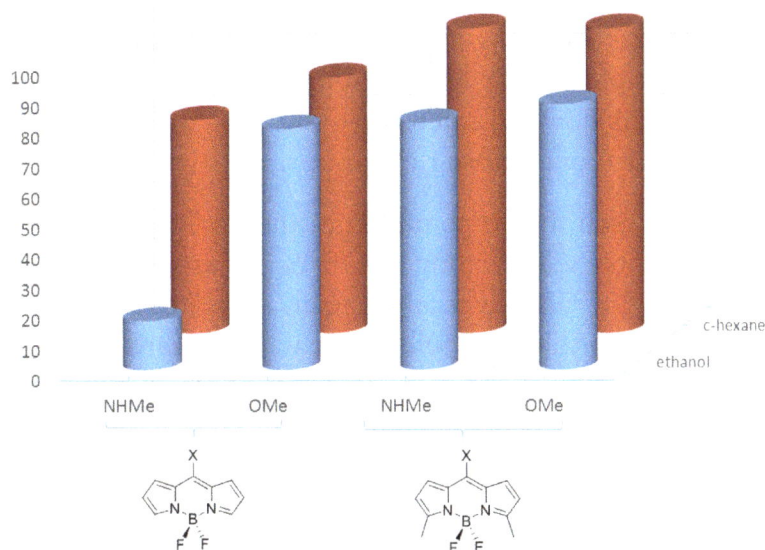

**Figure 5.** Fluorescent efficiencies of the BODIPYs bearing methylamine and methoxy at *meso* position in apolar (cyclohexane) and polar (ethanol) solvents. The corresponding data for their dimethylated derivatives at 3,5-positions are also included to highlight the ability of this alkylation to regulate the fluorescent response.

as to activate ICT since the fluorescent efficiency is always higher than 80% (**Figure 5**). Such ICT hampers the behavior of 8-aminoBODIPYs as fluorescent probe or laser dye. However, the simple methylation at key positions 3 and 5, hinders the population of the fluorescence quenching ICT and high fluorescent responses are recovered regardless of the media (always higher than 80%, and even approaching the 100% in apolar media, **Figure 5**) [29]. Such selective alkylation decreases the electron acceptor character of the BODIPY and reduces the probability of formation of ICT, allowing the recovering of a high fluorescent efficiency.

As a result, the laser performance of such dimethylated 8-heteroatomBODIPYs is excellent, not only in terms of efficiency but also in photostability [29]. We should bear in mind that one of the main drawbacks of commercially available blue-emitting dyes (such as coumarins, considered as the benchmark in this spectral region) is their low photostability under prolonged irradiation. The herein reported blue BODIPYs surpass the laser performance of these dyes maintaining the 45% and 70% of the laser emission after 40,000 pulses for the dyes bearing amine or methoxy, respectively, and reaching efficiencies up to 30% at 455 nm and 50% at 490 nm, respectively.

## 2.2. Red-emitting BODIPYs

The development of red-emitting BODIPYs is an active seeking subject. Fluorophores working in this spectral region are highly desirable and appealing for bioimaging purposes, owing to the ability of this kind of irradiation to penetrate deeper into tissues [4, 15]. Besides, the background autofluorescence of biomolecules (usually blue or green) is avoided, which is reflected in a more sensitive detection process. Most of the available dyes in the red spectral region are poorly fluorescent or has a low photostability. Thus, and in analogy to the strategy carried out in the preceding section, there is a great interest to shift the emission of BODIPYs to the red-edge, and thereby to circumvent in this way such photonic limitations taking advantage of the brightness and robustness of BODIPYs. In the bibliography, four major strategies have been reported to get the desired bathocromic shift [33]: (1) replacement of the central carbon at *meso* position by an aza group (aza-BODIPYs) [34]; (2) arylation of the dipyrrin core (fused BODIPYs) [35]; (3) extension of the aromatic π-system through aromatic framework; and (4) attachment of electron rich groups.

In this section, we have chosen the combination of the last two options as the strategy to develop bright and photostable red-emitting BODIPYs (**Figure 3**). To this aim, the aforementioned versatile 8-thiomethylBODIPY was taken again as scaffold. In this way, specific functionalization was added with orthogonal selectivity, which means that we can modulate the substitution patterns at positions 8, 2 and 6, and 3 and 5 independently, using non-interfering synthetic routes specific for each functionalization (nucleophilic substitution at *meso* position, Liebeskind-Srogl cross-coupling (LSCC) from easily accessible 2,6-halogenated compounds, and Knoevenagel condensation, from methylated 3,5-positions) [36]. By this synthetic approach, the dipyrrin core was decorated with a sterically hindered *ortho*-methylphenyl at *meso* position, a styryl arm bearing *para*-trifluoromethyl at positions 3 and 5, and finally a *para* substituted (with electron donor methoxy or withdrawing formyl) phenyl ring at positions 2 and 6 (**Figure 3**).

These fully functionalized dyes show absorption and emission bands placed deeply in the red/near infrared spectral window (absorption centered at around 660 nm, whereas fluorescence at around 700 nm, **Figure 6**). The frontier molecular orbitals reveal that the delocalized π-system is extended through the aromatic frameworks at positions 2 and 6, and mainly through positions 3 and 5, explaining the huge bathochromic shift recorded (around 160 nm with regard to the reference simplest BODIPY, **Figure 6**). Note that the constrained 8-aryl group does not take part in the delocalization since it is almost orthogonally disposed with regard to the dipyrrin plane. This an important fact in the molecular design, since it is well known that *meso*-aryl moieties with free motion switch on non-radiative deactivation channels, which drastically quench the fluorescence signal [37, 38]. The steric hindrance provided by the *ortho* methyl is enough to avoid such quenching pathway.

As a result of the rational design, high red fluorescent efficiencies are recorded in apolar media (reaching almost the 90% in cyclohexane, **Figure 7**). However, the evolution of the fluorescence quantum yield with the solvent polarity drastically depends on the functionalization at the 2,6-phenyl rings [36]. Thus, upon formylation of such aryl, the fluorescent efficiency slightly decreases in polar media (down to 73% in methanol, **Figure 7**). However, the replacement of such group by methoxy implies a clear decline on the fluorescent efficiency (down to 13%, **Figure 7**), together with a further bathochromic shift (emission placed at 710 nm, **Figure 6**). In the former compound, with the weak electron withdrawing formyl group, the trend is the expected one with BODIPYs; thus, this compound is an ideal candidate as fluorescent probe or laser dye in the red-edge. However, in the latter compound, bearing the strong electron donor methoxy, a quenching process is activated in polar media. Considering the push-pull character of the molecule (from 2,6-methoxy to 3,5-trifluoromethyl), an ICT is stabilized in polar media limiting the fluorescent response of the dye.

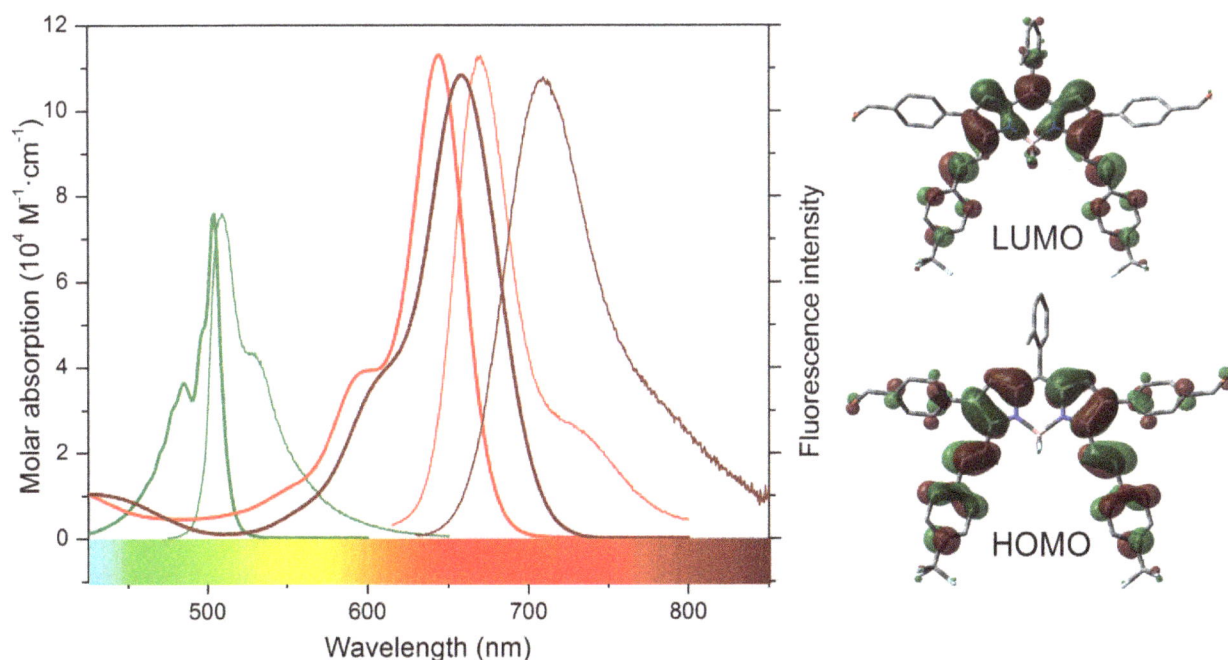

**Figure 6.** Absorption (bold line) and normalized fluorescence (thin line) spectra of the red-emitting dyes with regard to its reference unsubstituted counterpart in cyclohexane. The corresponding frontier contour maps of a representative red BODIPY is also enclosed.

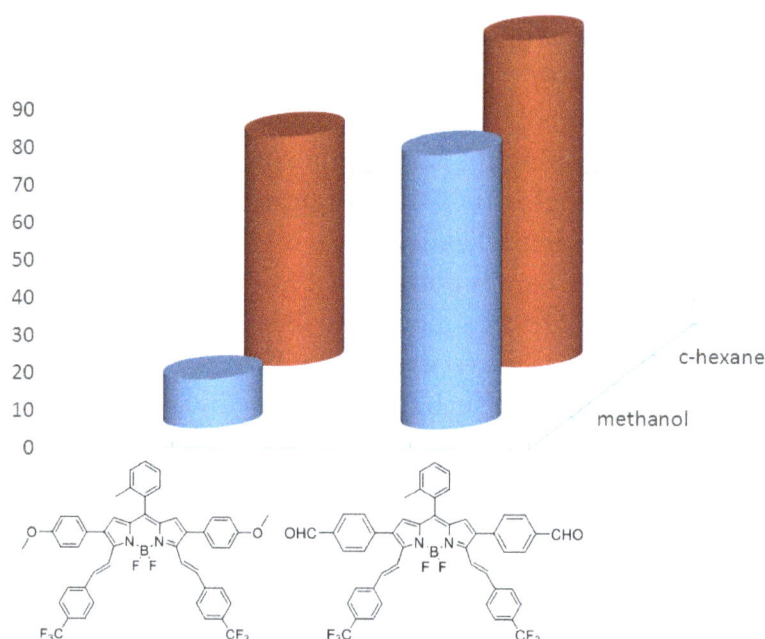

**Figure 7.** Fluorescent efficiencies of the red-emitting BODIPYs in apolar (cyclohexane) and polar (methanol) solvents.

Again, we tested the laser performance of these two dyes, and it was excellent [36]. The laser efficiencies reached the 60% in the formylated dye, being lower for the derivative bearing methoxy instead (43%, in agreement with the fluorescence measurements) and the laser emission is tunable from 700 to 740 nm, remaining high (90–100%) even after 50,000 pulses. Such performance is much better than that of stiryls (LDS dyes), which provide much lower efficiencies, or oxazines, which are more unstable. Therefore, these BODIPYs should behave nicely as red fluorescent probes or laser dyes.

Summing up, and taking into account the herein developed fluorophores in both sections (2.1 and 2.2), we have been able to translate the excellent photophysical signatures of BODIPYs to both sides of the visible spectral region, leading to fluorophores which can be considered standards or benchmark in terms of photonic behavior, such as fluorescent and lasing efficiency and photostability.

# 3. Induction of new photophysical processes

In the preceding sections, we have demonstrated the deep impact of the substitution pattern in the spectral shift. Moreover, in some cases, such functionalization was able to induce new photophysical phenomena such as ICT processes, which should be avoided to ensure high fluorescent response. Indeed, depending on the environmental conditions, it becomes the low-lying excited state and quenches efficiently the emission from the locally excited state. Therefore, usually dyes ongoing ICT are not suitable as fluorescent probes for bioimaging or active media of tunable dye lasers. However, and owing to the sensibility of the ICT to the surrounding environment properties, fluorophores undergoing ICT can be ideal as fluorescent sensors to monitor a specific characteristic of the media, following the fluorescent efficiency (on/off switches, if the ICT is not emissive) or the change in the fluorescence color (colorimetric sensors, if the ICT becomes fluorescence) [23–25]. In the next sections, we describe acidity/basicity and polarity sensors based on BODIPYs ongoing ICT processes (see molecular structures in **Figure 8**).

## 3.1. Sensor of acidity/basicity

In the previous section, we have shown the chemical versatility of the 8-methylthioBODIPY. For instance, in Section 2.1, the 8-heteroatomBODIPYs were attained by reaction of such scaffold with amine or alcohol and the ensuing displacement of the thiomethyl owing to the undergoing $S_N$Ar-like reaction. A similar protocol can be used to attach acetylacetonate (acac) groups to the said *meso* position of the BODIPY (**Figure 8**) [39]. These dyes substituted with active methylene groups offer a wide assortment of post-functionalization opportunities, since it can act as ligand for metals or semiconductors, and even itself can become a fluorophore after chelation of the acac with difluoroboron. The acac can exist in the keto or enol form, but in the herein reported compounds, theoretical calculations suggest that the tautomeric equilibrium is shifted to the enolic form, thereby providing an ionizable hydroxyl group.

The acac group has a scarce impact in the spectral band positions, being those typical of the BODIPY. Indeed, the optimized geometry predicts that the acac is disposed perpendicular to the dipyrrin plane avoiding an electronic coupling. As a consequence, the 8-acacBODIPY displays a bright green emission with fluorescent efficiencies approaching 100% (**Figure 9**). Such orthogonal disposition is a key feature to ensure high fluorescent deactivation. In fact, those BODIPY derivatives bearing $sp^2$ hybridized carbons linked at *meso* position (such as vinyl) are non-fluorescent owing to the undergoing planarity distortion in the dipyrrin plane upon selective electron coupling with one chromophoric pyrrole in the excited state [40–42]. One way to avoid such quenching is to induce high enough steric hindrance, as happens

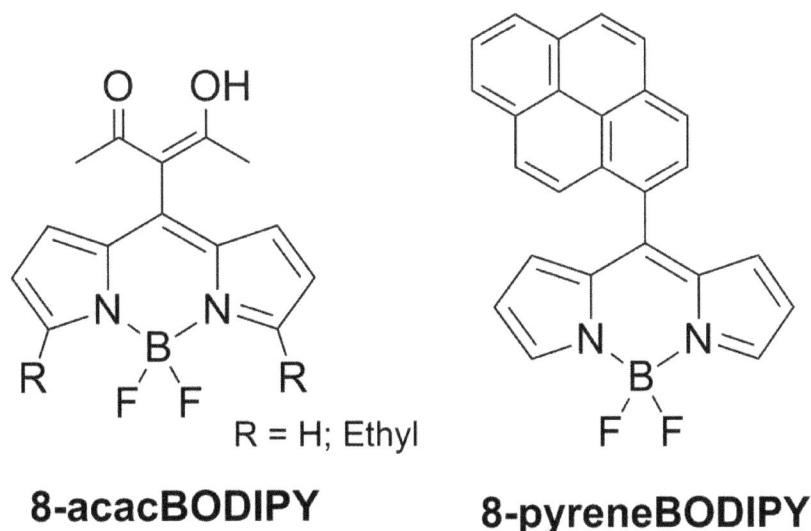

**8-acacBODIPY**          **8-pyreneBODIPY**

**Figure 8.** Molecular structure of the herein designed fluorescent sensors based on BODIPY.

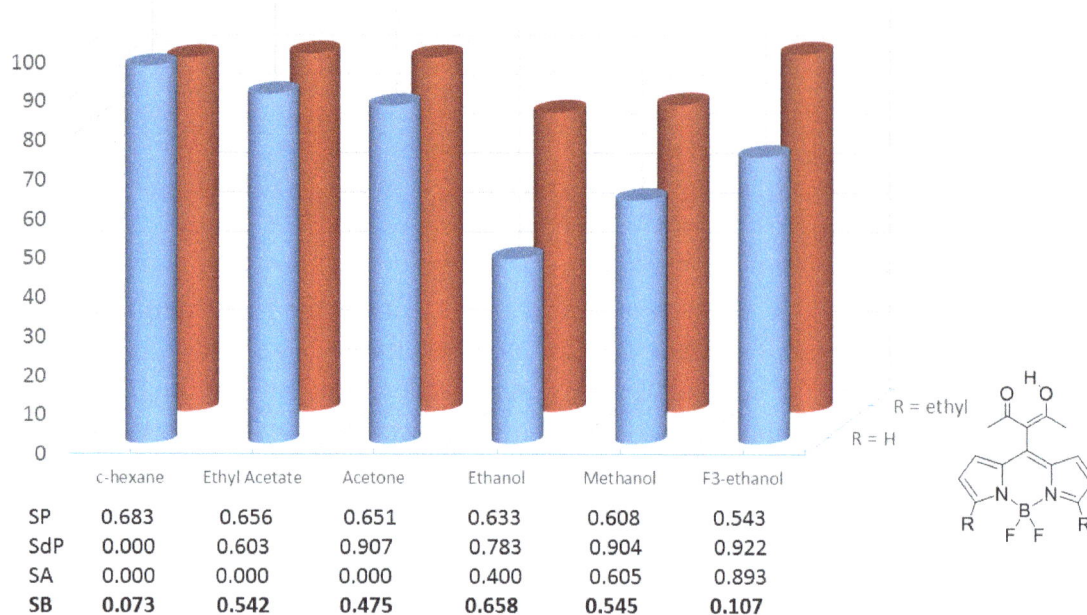

| | c-hexane | Ethyl Acetate | Acetone | Ethanol | Methanol | F3-ethanol |
|---|---|---|---|---|---|---|
| SP | 0.683 | 0.656 | 0.651 | 0.633 | 0.608 | 0.543 |
| SdP | 0.000 | 0.603 | 0.907 | 0.783 | 0.904 | 0.922 |
| SA | 0.000 | 0.000 | 0.000 | 0.400 | 0.605 | 0.893 |
| **SB** | **0.073** | **0.542** | **0.475** | **0.658** | **0.545** | **0.107** |

**Figure 9.** Evolution of the fluorescent efficiency of the 8-acacBODIPYs with the solvent properties described by the Catalan solvent scale (SP – Polarizability; SdP – Dipolarity; SA – Acidity; SB – Basicity).

with the acac. However, a closer inspection to the evolution of the fluorescent response with the solvent discloses that in alcohols, the efficiency drops more than expected (down to 47% in ethanol, **Figure 9**), whereas the lifetime remained high (around 7 ns). Using the solvent scales to account for the contribution of each solvent property (provided by Catalan) [43], we find that there is a correlation between the solvent basicity (SB scale) and the decrease in fluorescence (**Figure 9**). This is why we tested the influence of the pH of the media in the fluorescence signatures. Usually, the BODIPY dyes are rather hydrophobic and are not soluble in water. However, the presence of acac increases its hydrophilic character, and the dye becomes soluble in water, at least at low concentrations (around µM).

The fluorescence intensity decreases progressively as the pH of the media increases, thus at pHs more basic than 8 the fluorescence emission is almost completely lost (**Figure 10**). From the evolution of the fluorescence quantum yield with the pH a pKa of 6.9 is attained. Therefore, a quenching process takes place in basic media being reversible since acidification of basic solution implies a recovering of the fluorescence signal. Therefore, this dye behaves as an on/off switch to monitor the pH of the environment by measuring the evolution of the fluorescent response. Thus, at low pHs, the sensor is highly fluorescent (on), but at high pHs, the sensor is almost dark (off). Such sensing ability should be related with the ionization of the hydroxyl group of the enolic form of the 8-acac in basic conditions. Theoretical simulations show that the electronic distribution and energies of the HOMO and LUMO in acid media (acac as enol) is the typical of BODIPYs, thus being unaltered by attachment of the 8-acac (**Figure 10**). However, at basic media (pH higher than the pKa where the hydroxyl is ionized and the corresponding enolate is formed), the contour maps change markedly. The geometrical arrangement of the acac as enolate is no longer orthogonal with the dipyrrin, but it

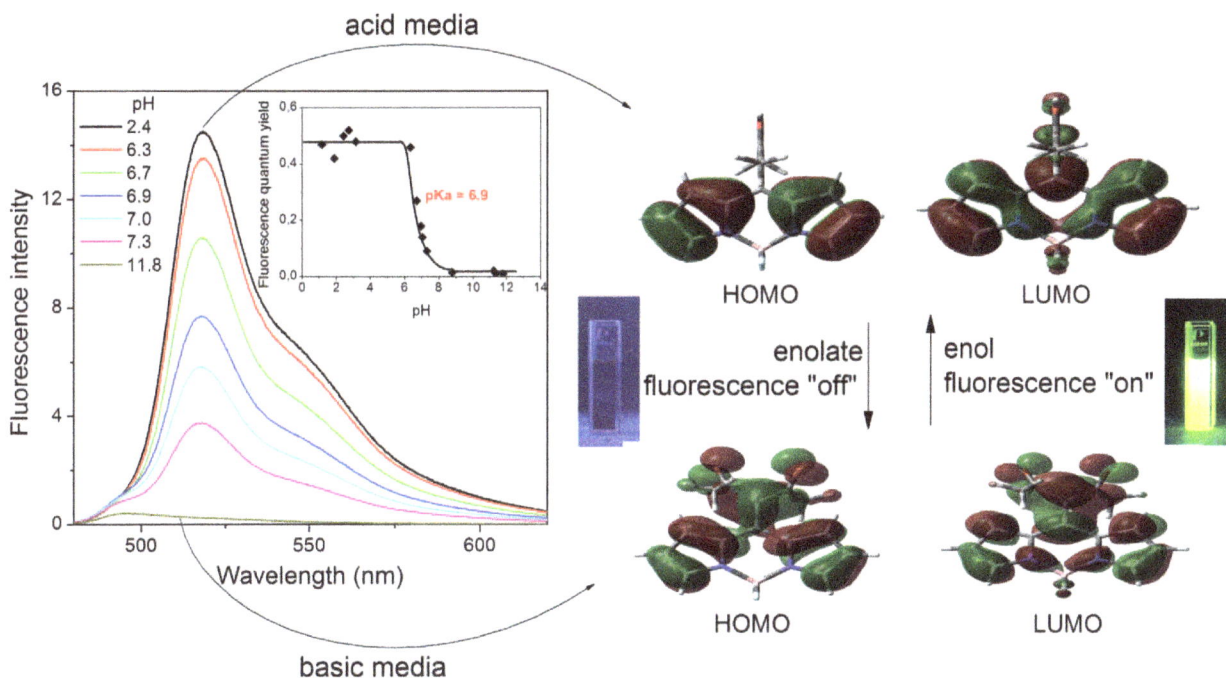

**Figure 10.** Evolution of the fluorescence band and quantum yield (inset) of the 8-acacBODIPY with the pH in aqueous solutions. The frontier contour maps before and after ionization of hydroxyl of the 8-acac are also enclosed.

becomes more coplanar (twisting dihedral angle around 30°), enabling an electron coupling. In this state, the negative charge of the enolate interact by resonant with the dipyrrin (see the electronic density spanned through the whole structure in both frontier orbitals), causing a planarity distortion in the chromophore (**Figure 10**). Therefore, the ICT from the ionized hydroxyl of the enolate to the BODIPY implies a drastic increase in the internal conversion probability and a loss of the aromaticity explaining the lack of fluorescence signal in basic media and allowing the sensing ability of this dye.

The effect of the ethylation of positions 3 and 5 is worthy of being outlined [39]. In agreement with the aforementioned results for 8-aminoBODIPYs in Section 2.1, the sole alkylation at these key positions enhances the fluorescent efficiency mainly in basic media (up to 76% in ethanol, **Figure 9**). It seems that the inductive donor effect of the ethyl groups infers a lower electron acceptor character to the BODIPY, being less predisposed to accept the electronic density from the 8-enolate, and hence lowering the quenching ICT pathway in basic media.

The laser measurements correlate well with the photophysical findings [39]. Indeed, the 8-acacBODIPY displays a green laser emission at 550 nm with an efficiency up to 55%, maintaining the 90% of the laser emission after 50,000 pulses. Ethanol is the solvent with the lower lasing efficiency (down to 43%) in agreement with the recorded decrease in the fluorescence quantum yield (**Figure 9**). Moreover, the 3,5-ethylation implies not only an increase in the laser efficiency (up to 70%, in concordance with the evolution of the fluorescent efficiency in **Figure 9**) but also it becomes more photostable (keeping the 100% of the emission under the same pumping regime). Therefore, these dyes bearing acac at *meso* position are rather versatile since they behave as laser dye and as on/off pH sensor as well.

## 3.2. Sensor of polarity

Another exploited strategy with BODIPYs is based on the development of multichromophoric dyes. In particular, that is related with the molecular design of excitation energy transfer (EET) cassettes [44]. The idea is to link covalently a suitable chromophore acting as energy donor to a complementary fluorophore playing the role of energy acceptor. This kind of dyads allows an efficient light harvesting over a broad spectral region, but provides almost exclusively the emission of the energy acceptor via a highly efficient intramolecular excitation energy transfer (intra-EET) owing to the short donor-acceptor distance imposed by the covalent linkage. The most common EET mechanism is mediated by the Förster formalism (FRET, also known as through-space), which demands a spectral overlap between the energy donor fluorescence band and the energy acceptor absorption spectrum to enable the required dipole-dipole coupling. Other reported EET mechanisms are Dexter (electronic exchange) or through-space (TBET, superexchange mechanism) [45, 46]. Following this approach, we tested the photophysical signatures of a dyad featuring a pyrene (blue emitting chromophore) linked to the *meso* position of the BODIPY (green emitting fluorophore) core (**Figure 8**). The pyrene was chosen since it can be grafted to the *meso* position of the 8-methylthioBODIPY platform via a LSCC reaction from the adequate boronic acid [47]. Besides its fluorescence profile, it is suitable to undergo FRET with the BODIPY since the required spectral overlap is feasible [48].

The absorption profile of the dyad shows the characteristic bands of both chromophores (**Figure 11A**). Thus, the typical vibrational resolution of the electronic band of the pyrene is recorded in the UV, followed by the sharp absorption of the BODIPY at the Vis region. Therefore, both chromophores contribute additively to the overall broadband absorption covering the UV-blue-green spectral region. In fact, the electron coupling between both moieties is disabled in the ground state since the pyrene is twisted around 65° due to steric reasons [48]. Regardless of the excitation wavelength (directly at the BODIPY Vis absorption or at the pyrene UV absorption), the Vis emission outcoming from the BODIPY prevails in the fluorescence profile, whereas the emission from the pyrene is strongly quenched (**Figure 11A**). This experimental finding supports that the dyad undergoes intra-EET from the donor pyrene to the BODIPY acceptor and final light emitting moiety. Likely, the ongoing mechanism is FRET due to the available spectral overlap, but also TBET could be viable since both chromophores are directly linked enabling the required electronic exchange [45, 46]. The fluorescent efficiency (30% in apolar cyclohexane) and lifetime (around 2 ns) of the dyad are lower than the expected in BODIPYs. This moderate fluorescence loss can be attributed to the free motion of the bulky pyrene at the key and substituent sensitive *meso* position, which enhances the internal conversion probability.

Nonetheless, the above described fluorescence signatures in apolar media dramatically change in polar media (**Figure 11B**). Whereas no changes are detected in the absorption features, the fluorescence profile is deeply altered. The emission at 520 nm (assigned to the locally excited (LE) state) is strongly quenched, and a new broad and red-shifted emission is recorded (around 640 nm and reaching 700 nm in the most polar media). Besides, the fluorescent efficiency dramatically decreases in the more polar media where such new emission becomes rather weak and almost negligible. These experimental findings reveal complex

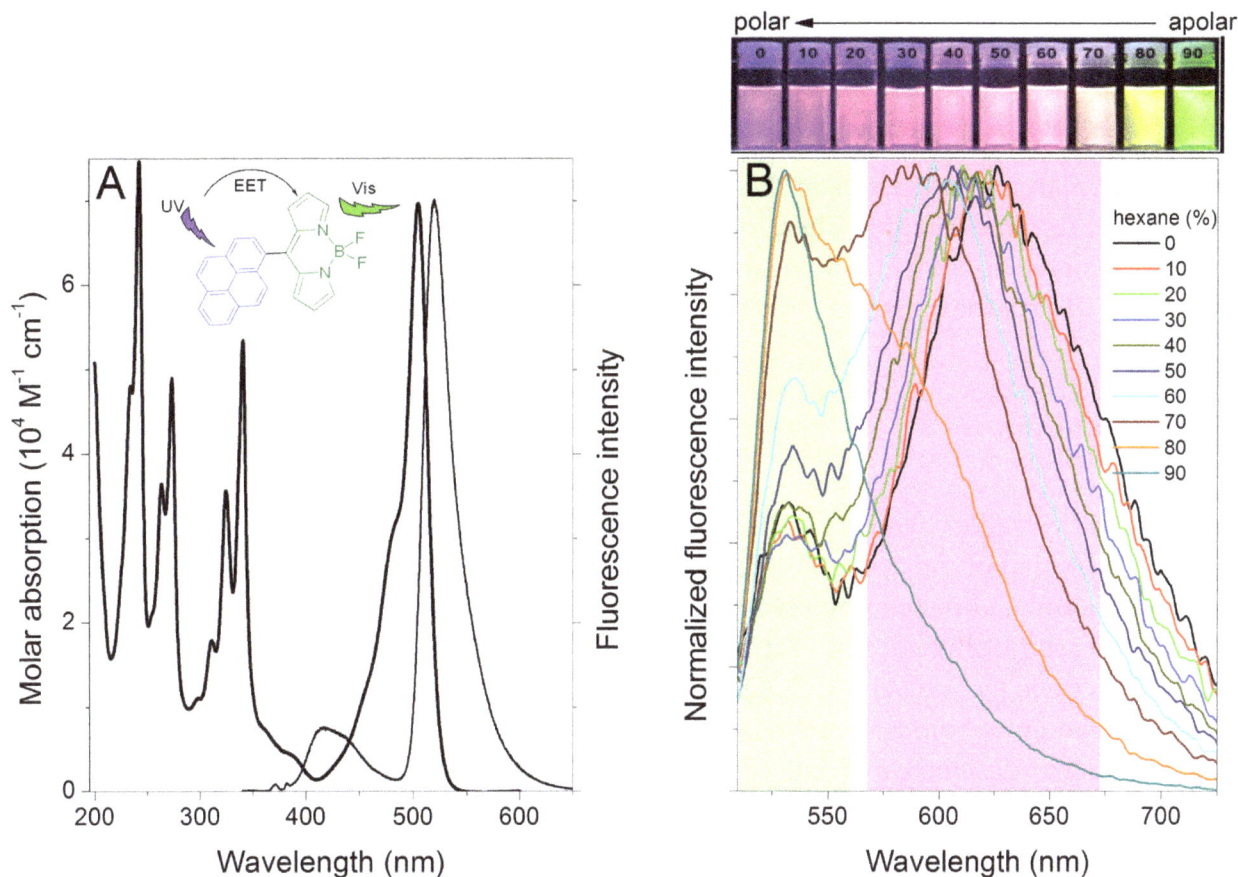

**Figure 11.** (A) UV–Vis absorption (bold line) and fluorescence (thin line, upon UV excitation at 340 nm) spectra of the 8-pyreneBODIPY in cyclohexane. A sketch of the ongoing energy transfer is also enclosed. (B) Evolution of the fluorescence profile with the solvent polarity (normalized spectra in hexane-tetrahydrofuran mixtures). The two detection channels and the corresponding fluorescence images in each mixture proportion are included to highlight the sensing ability.

molecular dynamics in the excited state (EET still takes place in polar media) since a new low-lying emitting state in formed upon excitation in polar media. All these trends pinpoint to the activation of an ICT state from the pyrene to the BODIPY. The high dipole moment of the ICT owing to its inherent charge separation explain why is further stabilized in polar media. In the preceding sections, a non-fluorescence ICT was already claimed to explain the recorded fluorescence quenching in polar media. However, in this case, the ICT becomes fluorescence displaying a long-wavelength emission very sensitive to the solvent polarity. An increase of the polarity of the media implies a more efficient quenching of the emission from the LE state, and also from the ICT as well, since the charge recombination probability is lower and the charge separation is further favored [49, 50].

Although this dyad is not valuable as laser dye or fluorescent probe, it can be applied as polarity sensor of the surrounding environment owing to the ICT formation and its high sensitivity to this solvent property. As a matter of fact, we have systematically varied the medium polarity using mixtures of apolar solvent (hexane) and a polar one (tetrahydrofuran) (**Figure 11B**). Thus, in apolar media, just a green emission is recorded from the LE state. However, a progressive increase of the solvent polarity implies that the emission from the ICT is detected in detriment of that from the LE state. Thus, in polar environment, both emissions are simultaneously detected, leading to a yellow emission. Finally, in the more polar media, the emission from the ICT exceeds that from the LE and prevails, leading to a purple emission. Summing up, in this dyad, where the ICT fluorescence deactivation is feasible, two channels for detection are provided; one at the green part of the visible (around 520 nm) and other at the red part (around 640 nm). Therefore, the solvent polarity can be monitored by measuring the change of the fluorescence intensity at these wavelength, or alternatively just by the naked eye by visualizing the emission color under UV excitation owing to the ongoing intra-EET.

# 4. Conclusions

The rich and versatile chemistry of the robust boron-dipyrrin core allows the design and synthesis of a myriad of molecular structures with diverse photophysical signatures, which can be finely modulated depending on the target application field. This tailoring ability is likely the main reason of the success and reputation of these chameleonic fluorophores. It is not risky to envisage that nowadays BODIPYs are the most recommended luminophore for opto-electronics and biophotonics, or whatever utility in which an organic dye could be involved. After a rational design, a molecule fitting the requirements of a specific application can be customized, guaranteeing a nice and accurate performance.

In this chapter, we settle some structural guidelines to orient the synthesis of novel BODIPYs as candidates to be applied as fluorescent probes, laser dyes or fluorescent sensors along the entire visible spectral region. All the herein tested BODIPY derivatives are attained from the 8-methylthioBODIPY scaffold. This versatile platform is able to undergo different synthetic routes selectively in orthogonal positions, in particular at *meso* position, at $\beta$ (2 and 6) and $\alpha$ (3 and 5) pyrrolic positions. Such reactivity enables a rich and exhaustive functionalization of the dipyrrin core.

One of the main pursue goals was to translate the excellent stability and fluorescent response of BODIPYs working usually in the middle part of the visible spectrum (green-yellow) to both edges (blue and red), giving rise to standard and optimal fluorophores across this spectral region. On the one hand, to push the spectral bands toward the red-edge aromatic frameworks bearing electron rich groups were grafted to the BODIPY core, mainly at positions 3 and 5. Push-pull chromophores (bearing electron donor and acceptor functionalities) provide more pronounced red spectral shift, but the fluorescence emission is damaged owing to the induced charge separation (ICT). On the other hand, the opposite blue shift was achieved changing the electronegativity of the heteroatom attached at the *meso* position. The electron coupling of these heteroatoms with the dipyrrin switches on new delocalized $\pi$-system responsible of the hypsochromic shift. To ensure high fluorescent efficiencies, it is recommended to alkylate the key positions 3 and 5.

Taking into account that the *meso* position is the most sensitive one of the backbone to the substituent effect, it was selected to attach functionalities able to induce ICT processes. These phenomena should be avoided if a high fluorescent response is required, but it is ideal to develop sensors due to its sensitive to specific environmental properties. On the one hand, the attachment of active methylene at such position leads to bright green dyes. However, such emission is lost in basic media. Such reversible quenching allows the monitorization of the pH of the surrounding environment following the change of the green fluorescence intensity (on/off switch). On the other hand, the grafting of pyrene enables excitation far away from the emission region owing to energy transfer processes. At the same time, such functionalization induces an ICT state characterized by red emission. The ratio between the green and red fluorescence is very sensitive to the solvent polarity. Thus, this sensor allows to visualize the polarity of the environment just by the color of fluorescence.

Although the number of reports dealing with BODIPYs is huge and still growing fast, we foresee that the future of these dyes is far for being over, and still plenty of applications need to be explored with this marvelous and amazing organic luminophore.

# Acknowledgements

Financial support from MICINN (MAT2017-83856-C3-3-P), Gobierno Vasco (IT912-16), and CONACyT (Mexico, Grant 253623) are acknowledged. R. S-L. and E.A-Z. thank Gobierno Vasco for a postdoctoral contract and predoctoral fellowship, respectively. C. F. A. G.-D. and J. L. B.-V. thank CONACyT for doctoral scholarship.

# Author details

Rebeca Sola Llano[1], Edurne Avellanal Zaballa[1], Jorge Bañuelos[1]*,
César Fernando Azael Gómez Durán[2], José Luis Belmonte Vázquez[2], Eduardo Peña Cabrera[2] and Iñigo López Arbeloa[1]

*Address all correspondence to: jorge.banuelos@ehu.eus

1 Department of Physical Chemistry, University of the Basque Country (UPV/EHU), Bilbao, Spain

2 Department of Chemistry, University of Guanajuato, Guanajuato, Mexico

# References

[1] Betzig E. Single molecules, cells, and super-resolution optics (nobel lecture). Angewandte Chemie, International Edition. 2015;**54**:8034-8053. DOI: 10.1002/anie.201501003

[2] Hell SW. Nanoscopy with focused light (nobel lecture). Angewandte Chemie, International Edition. 2015;**54**:8054-8066. DOI: 10.1002/anie.201504181

[3] Moerner WE. Single-molecule spectroscopy, imaging, and photocontrol: Foundations

for super-resolution microscopy (nobel lecture). Angewandte Chemie, International Edition. 2015;**54**:8067-8093. DOI: 10.1002/anie.201501949

[4]   Sinkeldam RW, Greco NJ, Tor Y. Fluorescent analogs of biomolecular building blocks: Design, properties, and applications. Chemical Reviews. 2010;**110**:2579-2619. DOI: 10.1021/cr900301e

[5]   Chénais S, Forget S. Recent advances in solid-state organic lasers. Polymer International. 2012;**61**:390-406. DOI: 10.1002/pi.3173

[6]   Kuehne AJC, Gather M. Organic lasers: Recent developments on materials, device geometries and fabrication techniques. Chemical Reviews. 2016;**116**:12823-12864. DOI: 10.1021/acs.chemrev.6b00172

[7]   Moliner F, Kielland N, Lavilla R, Vendrell M. Modern synthetic avenues for the preparation of functional fluorophores. Angewandte Chemie, International Edition. 2017;**56**:3758-3769. DOI: 10.1002/anie.201609394

[8]   Benniston AC, Copley G. Lighting the way ahead with boron dypiromethene (Bodipy) dyes. Physical Chemistry Chemical Physics. 2009;**11**:4121-4131. DOI: 10.1039/b901383k

[9]   Bañuelos J. BODIPY dye, the most versatile fluorophore ever? Chemical Record. 2016;**16**:335-348. DOI: 10.1002/tcr.201500238

[10]  Loudet A, Burgess K. BODIPY dyes and their derivatives: Syntheses and spectroscopic properties. Chemical Reviews. 2007;**107**:4891-4932. DOI: 10.1021/cr078381n

[11]  Benstead M, Mehl GH, Boyle RW. 4,4'-Difluoro-4-bora-3a,4a-diaza-s-indacenes (BODIPYs) as components of novel light active materials. Tetrahedron. 2011;**67**:3573-3601. DOI: 10.1016/j.tet.2011.03.028

[12]  Ulrich G, Ziessel R, Harriman A. The chemistry of fluorescent Bodipy dyes: Versatility unsurpassed. Angewandte Chemie, International Edition. 2008;**47**:1184-1201. DOI: 10.1002/anie.200702070

[13]  Boens N, Verbelen B, Dehaen W. Postfunctionalization of the BODIPY core: Synthesis and spectroscopy. European Journal of Organic Chemistry. 2015:6577-3595. DOI: 10.1002/ejoc.201500682

[14]  Lakshmi V, Sharma R, Ravikanth M. Functionalized boron-dipyrromethenes and their applicattions. Reports in Organic Chemistry. 2016;**6**:1-24. DOI: 10.2147/ROC.S60504

[15]  Kowada T, Maeda H, Kikuchi K. BODIPY-based probes for the fluorescence imaging of biomolecules in living cells. Chemical Society Reviews. 2015;**44**:4953-4972. DOI: 10.1039/c5cs00030k

[16]  Costela A, García-Moreno I, Sastre R. Polymeric solid-state dye lasers: Recent developments. Physical Chemistry Chemical Physics. 2003;**5**:4745-4763. DOI: 10.1039/b307700b

[17]  Bessette A, Hanan GS. Desig, synthesis and photophysical studies of dipyrromethene-based materials: Insights into their applications in organic photovoltaic devices. Chemical Society Reviews. 2014;**43**:3342-3405. DOI: 10.1039/c3cs60411j

[18] Singh SP, Gayathri T. Evolution of BODIPY dyes as potential sensitizers for dye-sensitized solar cells. European Journal of Organic Chemistry. 2014:4689-4707. DOI: 10.1002/ejoc.201400093

[19] Nepomnyashchii AB, Bard AJ. Electrochemistry and electrogenerated chemiluminescence of BODIPY dyes. Accounts of Chemical Research. 2012;**45**:1844-1853. DOI: 10.1021/a200278b

[20] Boens N, Leen V, Dehaen W. Fluorescent indicators based on BODIPY. Chemical Society Reviews. 2012;**41**:1130-1172. DOI: 10.1039/c1cs15132k

[21] Zhao J, Wu W, Sun J, Guo S. Triplet photosensitizers: From molecular design to applications. Chemical Society Reviews. 2013;**42**:5323-5351. DOI: 10.1039/c3cs35531d

[22] Kamkaew A, Lim SH, Lee HB, Kiew LV, Chung LY, Burgess K. BODIPY dyes in photodynamic therapy. Chemical Society Reviews. 2013;**42**:77-88. DOI: 10.1039/c2cs35216h

[23] Prasanna de Silva A, Gunaratne HQN, Gunnlaugsson T, Huxley AJM, McCoy CP, Rademacher JT, Rice TE. Signaling recognition events with fluorescent sensors and switches. Chemical Reviews. 1997;**97**:1515-1566. DOI: 10.1021/cr960386p

[24] Valeur B, Leray I. Design principles of fluorescent molecular sensors for cation recognition. Coordination Chemistry Reviews. 2000;**205**:3-40. DOI: 10.1016/S0010-8545(00)00246-0

[25] Rurack K, Resch-Genger U. Rigidization, preorientation and electronic decoupling – The `magic triangle' for the design of highly efficient fluorescent sensors and switches. Chemical Society Reviews. 2002;**31**:116-127. DOI: 10.1039/b100604p

[26] Pavlopoulos TG. Scaling of laser dyes with improved laser dyes. Progress in Quantum Electronics. 2002;**26**:193-224. DOI: 10.1016/S0079-6727(02)00005-8

[27] López Arbeloa F, Bañuelos J, Martínez V, Arbeloa T, López Arbeloa I. Structural, photophysical and lasing properties of pyrromethene dyes. International Reviews in Physical Chemistry. 2005;**24**:339-373. DOI: 10.1080/01442350500270551

[28] Arroyo IJ, Hu R, Merino G, Zhong Tang B, Peña-Cabrera E. The smallest and one of the brightest. Efficient preparation and optical description of the parent borondipyrromethene system. The Journal of Organic Chemistry. 2009;**74**:5719-5722. DOI: 10.1021/jo901014w

[29] Esnal I, Valois-Escamilla I, Gómez-Durán CFA, Urías-Benavides A, Betancourt-Mendiola ML, López-Arbeloa I, Bañuelos J, García-Moreno I, Costela A, Peña-Cabrera E. Chemphyschem. 2013;**14**:4134-4142. DOI: 10.1002/cphc.201300818

[30] Sathyamoorthi G, Soong ML, Ross TW, Boyer JH. Fluorescent tricyclic β-azavinamidine-BF$_2$ complexes. Heteroatom Chemistry. 1993;**4**:603-608. DOI: 10.1002/hc.520040613

[31] Bañuelos J, Lópz Arbeloa F, Martinez V, Liras M, Costela A, García Moreno I, López Arbeloa I. Difluoro-boron-triaza-anthracene: A laser dye in the blue region. Theoretical simulation of alternative difluoro-boron-diaza-aromatic systems. Physical Chemistry Chemical Physics. 2011;**13**:3437-3445. DOI: 10.1039/c0cp01147a

[32]  Goud TV, Tutar A, Biellmann JF. Synthesis of 8-heteroatom-substituted 4,4-difluoro-4-bora-3a,4a-diaza-s-indacene dyes (BODIPY). Tetrahedron. 2006;**62**:5084-5091. DOI: 10.1016/j.tet.2006.03.036

[33]  Lu H, Mack J, Yang Y, Shen Z. Structural modification strategies for the rational design of red/NIR region BODIPYs. Chemical Society Reviews. 2014;**43**:4778-4823. DOI: 10.1039/c4cs00030g

[34]  Ge Y, ÓShea DF. Azadipyrromethenes: From traditional dye chemistry to leading edge applications. Chemical Society Reviews. 2016;**45**:3846-3864. DOI: 10.1039/c6cs00200e

[35]  Umezawa K, Matsui A, Nakamura Y, Citterio D, Suzuki K. Brigth, color-tunable fluorescent dyes in the Vis/NIR region: Establishment of new "tailor-made" multi-color fluorophores based on borondipyrromethene. Chemistry - A European Journal. 2009;**15**:1096-1106. DOI: 10.1002/chem.200801906

[36]  Gómez-Durán CFA, Esnal I, Valois-Escamilla I, Urías-Benavides A, Bañuelos J, López-Arbeloa I, García-Moreno I, Peña-Cabrera E. Near-IR BODIPY dyes à la carte – Programmed orthogonal functionalization of rationally designed building blocks. Chemistry - A European Journal. 2016;**22**:1048-1061. DOI: 10.1002/chem.201503090

[37]  Kee HL, Kirmaier C, Yu L, Thamyongkit P, Youngblood WJ, Calder ME, Ramos L, Noll BC, Bocian DF, Scheidt WR, Birge RR, Lindsey JS, Holten D. Structural control of the photodynamics of boron-dipyrrin complexes. The Journal of Physical Chemistry. B. 2005;**109**:20433-20443. DOI: 10.1021/jp0525078

[38]  Prlj A, Vannay L, Corminboeuf C. Fluorescence quenching in BODIPY dyes; the role of intramolecular interactions and charge transfer. Helvetica Chimica Acta. 2017;**100**:e1700093. DOI: 10.1002/hlca.201700093

[39]  Gutierrez-Ramos B, Bañuelos J, Arbeloa T, López-Arbeloa I, Gónzalez-Navarro PE, Wrobel K, Cerdán L, García-Moreno I, Costela A, Peña-Cabrera E. Straightforward synthetic protocol for the introduction of stabilized C nucleophiles in the BODIPY core for advanced sensing and photonic applications. Chemistry - A European Journal. 2015;**21**:1755-1764. DOI: 10.1002/chem.201405233

[40]  Chibani S, Le Guennic B, Charaf-Eddin A, Laurent AD, Jacquemin D. Revisiting the optical signatures of BODIPY with ab initio tools. Chemical Science. 2013;**4**:1950-1693. DOI: 10.1039/c3sc22265a

[41]  Lincoln R, Greene LE, Bain C, Flores-Rizo JO, Bohle DS, Cosa G. When push comes to shove: Unravelling the mechanism and scope of nonemissive meso-unsaturated BODIPY dyes. The Journal of Physical Chemistry. B. 2015;**119**:4758-4765. DOI: 10.1021/acs.jpcb.5b02080

[42]  Prlj A, Fabrizio A, Corminboeuf C. Rationalizing fluorescence quenching in meso-BODIPY dyes. Physical Chemistry Chemical Physics. 2016;**18**:32668-32672. DOI: 10.1039/c6cp06799a

[43]  Catalán J. Toward a generalized treatment of the solvent effect based on four empirical

scales: Dipolarity (SdP, a new scale), polarizability (SP), acidity (SA) and basicity (SB) of the medium. The Journal of Physical Chemistry. B. 2009;**113**:5951-5960. DOI: 10.1021/jp8095727

[44] Fan J, Hu M, Zhang P, Peng X. Energy transfer cassettes based on organic fluorophores: Construction and applications in ratiometric sensing. Chemical Society Reviews. 2013;**42**:29-43. DOI: 10.1039/c2cs35273g

[45] Speiser S. Photophysics and mechanisms of intramolecular electronic energy transfer in bichromophoric molecular systems: Solution and supersonic jet studies. Chemical Reviews. 1996;**96**:1953-1976. DOI: 10.1021/cr941193+

[46] Curuchet C, Feist FA, Van Averbeke B, Mennuci B, Jacob J, Müllen K, Basché T, Beljonne D. Superexchange-mediated electronic energy transfer in a model dyad. Physical Chemistry Chemical Physics. 2010;**12**:7378-7385. DOI: 10.1039/c003496g

[47] Peña-Cabrera E, Aguilar-Aguilar A, González-Domínguez M, Lager E, Zamudio-Vázquez R, Godoy-Vargas J, Villanueva-García F. Simple, general, and efficient synthesis of meso-substituted borondipyrromethenes from a single platform. Organic Letters. 2007;**9**:3985-3988. DOI: 10.1021/ol7016615

[48] Bañuelos J, Arroyo-Córdoba IJ, Valois-Escamilla I, Alvarez-Hernández A, Peña-Cabrera E, Hu R, Zhong Tang B, Esnal I, Martínez V, López Arbeloa I. Modulation of the photophysical properties of BODIPY dyes by substitution at their meso position. RSC Advances. 2011;**1**:677-684. DOI: 10.1039/c1ra00020a

[49] Benniston AC, Clift S, Hagon J, Lemmetyinen H, Tkachenko NV, Clegg W, Harrington RW. Effect on charge transfer and charge recombination by insertion of a naphthalene-based bridge in molecular dyads based on borondipyrromethene (Bodipy). Chemphyschem. 2012;**13**:3672-3681. DOI: 10.1002/cphc.201200510

[50] Nano A, Ziessel R, Stachelek P, Harriman A. Charge-recombination fluorescence from push-pull electronic systems constructed around amino-substituted styryl-BODIPY dyes. Chemistry - A European Journal. 2013;**19**:13528-13537. DOI: 10.1002/chem.201301045

# 7

# Ultrafast Intramolecular Proton Transfer Reaction of 1,2-Dihydroxyanthraquinone in the Excited State

Sebok Lee, Myungsam Jen, Kooknam Jeon, Jaebeom Lee, Joonwoo Kim and Yoonsoo Pang

**Abstract**

1,2-Dihydroxyanthraquinone (alizarin) shows an ultrafast intramolecular proton transfer in the excited states between the adjacent hydroxyl and carbonyl groups. Due to the ground and electronic structure of locally excited and proton-transferred tautomers, alizarin shows dual emission bands with strong Stokes shifts. The energy barriers between the locally excited (LE) and proton-transferred (PT) tautomers in the excited state are strongly dependent on the solvent polarity and thus alizarin shows complicated photophysical properties including solvent and excitation dependences. The excited-state intramolecular proton transfer (ESIPT) of alizarin was monitored in time-resolved stimulated Raman spectroscopic investigation, where the instantaneous structural changes of anthraquinone backbone in 70~80 fs were captured. Two major vibrational modes of alizarin, $\nu$(C=C) and $\nu$(C=O) represent the proton transfer reaction in the excited state, which then leads to the vibrational relaxation of the product and the restructuring of solvent molecules. Ultrafast changes in solvent vibrational modes of dimethyl sulfoxide (DMSO) were also investigated for the solvation dynamics including hydrogen bond breaking and reformation.

**Keywords:** excited-state intramolecular proton transfer, tautomerization, femtosecond stimulated Raman, transient absorption

## 1. Introduction

Proton transfer occurring either intramolecularly or intermolecularly is one of the fundamental chemical reactions and has been of great interest in chemistry, biology, and related disciplines [1–5]. Molecules with the excited-state intramolecular proton transfer (ESIPT) often show large Stokes shifts, which is beneficial in many photonic applications due to small self-absorption [6, 7]. The ESIPT reactions have been extensively studied by time-resolved spectroscopic methods, where the ultrafast laser pulses initiate the chemical reaction in the excited state [6–10]. Femtosecond transient absorption technique was used as the time-resolved electronic probe in monitoring ultrafast proton transfer reactions in the time scales of ~30 fs [8, 10]. Recently, a much faster ESIPT of ~13 fs in 10-hydroxybenzo[*h*]quinolone has been reported by a fluorescence upconversion technique [9].

1,2-Dihydroxyanthraquinone (alizarin) is one of natural red pigments which forms an intra-molecular hydrogen bond between a carbonyl and a hydroxyl group in the ground and excited states [11–18]. Upon photoexcitation, a proton transfer from the hydroxyl to the carbonyl group occurs and the dual emission bands of the locally excited (LE) and proton-transferred (PT) tautomers have been reported [11–14]. The scheme of electronic structure of alizarin in LE and PT tautomers is shown in **Figure 1**. Since the barrier between LE and PT tautomers of alizarin in the excited state is tunable by changing solvent polarity [19, 20], the proton transfer reaction dynamics between the LE and PT tautomers might also be controlled by this factor. The emission band of the PT tautomer dominates in nonpolar aprotic solvents while the dual emission bands both from the LE and PT tautomers appear in polar aprotic solvents with the inhibition of the ESIPT reaction [21, 22]. According to the density functional theory (DFT)/time-dependent DFT (TDDFT) simulation results, the LE tautomer of alizarin is more stable (4.7–4.8 kcal/mol) than the PT tautomer in ground state, while the LE tautomer becomes

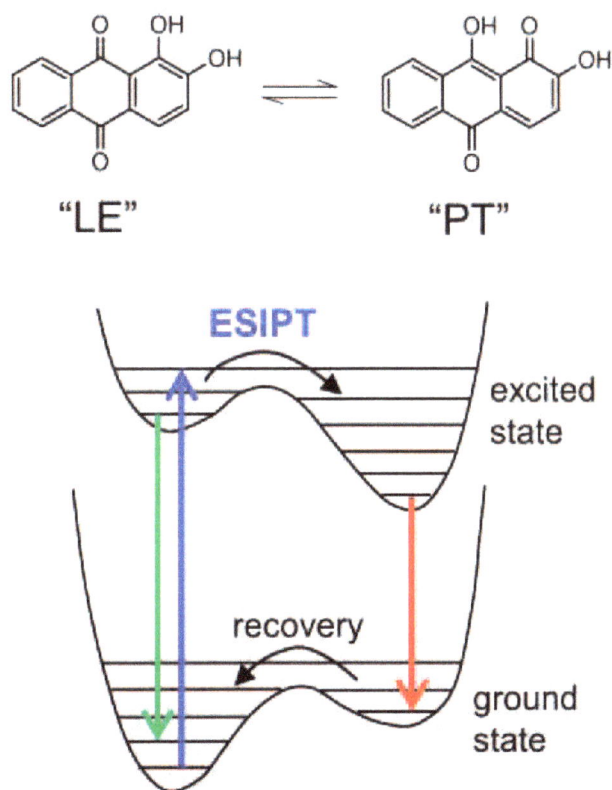

**Figure 1.** Electronic structure of alizarin in LE and PT tautomers. Adapted with permission from reference [41]. Copyright 2015 Elsevier.

less stable in the excited state with tunable energy barriers between the LE and PT [23]. For example, the energy barrier from the LE to PT tautomer in the exited state was estimated as 1.30 kcal/mol in benzene (nonpolar aprotic) and 3.19 kcal/mol in ethanol (polar aprotic).

The excited state lifetime of the alizarin PT tautomer was observed as 60–80 ps in several time-resolved spectroscopic investigations on alizarin, but the exact proton transfer dynamics of alizarin was not clearly obtained due to strong excited state absorption and emission signals, and complicated excited state dynamics including vibrational relaxations, solvations, etc., [20, 24–26]. However, the ultrafast ESIPT reactions (45–120 fs) of several anthraquinone derivatives including 1-hydroxyanthraquinone and 1-chloroacetylaminoanthraquinone have been measured by time-resolved fluorescent measurements [11, 15, 27, 28].

Femtosecond stimulated Raman spectroscopy (FSRS) with both high temporal (<50 fs) and spectral (<10 cm$^{-1}$) resolutions was introduced recently for the study of excited state dynamics and reaction mechanisms [29, 30] and has been widely used to study the photo-induced population and structural dynamics in many chemical and biological systems [31–34].

In this chapter, the ESIPT reaction and excited state dynamics of alizarin will be overviewed by using experimental results of steady-state absorption and emission, femtosecond transient absorption, and femtosecond stimulated Raman measurement. The excited state dynamics of alizarin was examined by changing the solvent polarity and the evidence for the ultrafast proton transfer reaction and subsequent structural changes in the product state were inspected by the time-dependent skeletal vibrational modes of alizarin.

## 2. Experimental details

### 2.1. Chemical preparation

Alizarin (Sigma-Aldrich, St. Louis, MO), dimethyl sulfoxide (DMSO, Daejung Chemicals and Metals, Siheung, Korea), ethanol (Duksan Pure Chemicals, Ansan, Korea), and other chemicals were used as received. Alizarin hardly dissolves in water but dissolves in most organic solvents, so alizarin solutions (33–50 μM) were prepared in ethanol and DMSO for the steady-state absorption and emission, and transient absorption measurements. A 2 mm cell with a stirring magnet was used for transient absorption measurements to avoid photo-damage from the laser pulses. The DMSO solutions of alizarin up to 20-mM concentrations in a 0.5 mm flow cell recirculated by a peristaltic pump were used for stimulated Raman measurements.

### 2.2. Steady-state absorption and emission measurements

The absorption spectra were recorded by a UV/Vis spectrometer (S-3100, Scinco, Seoul, Korea) and the emission spectra were obtained by a time-resolved fluorescence setup based on a time-correlated single photon-counting module (Picoharp 300, PicoQuant, Berlin, Germany), a pico-second diode laser ($\lambda_{ex}$ = 405 nm; P-C-405, PicoQuant), a monochromator (Cornerstone 260, Newport Corp., Irvine, CA), and a photomultiplier tube detector (PMA 192, PicoQuant).

## 2.3. Transient absorption spectroscopy

A femtosecond transient absorption setup based on a Ti:sapphire regenerative amplifier (LIBRA-USP-HE, Coherent Inc., Santa Clara, CA) was used for transient absorption measurements [35, 36]. The pump pulses at 403 nm were generated by sum-harmonic generation (SHG) in a BBO crystal ($\theta = 29.2°$, Eksma Optics, Vilnius, Lithuania) and compressed in a prism-pair compressor. The whitelight supercontinuum probe pulses (450–1000 nm) generated in a sapphire window were tightly focused to the sample with the pump and detected with a fiber-based spectrometer (QE65Pro, Ocean Optics, Largo, FL). Transient absorption spectra and kinetics were analyzed in a global fit analysis by using a software package Glotaran [37].

## 2.4. Femtosecond stimulated Raman spectroscopy

A femtosecond stimulated Raman setup based on the Ti:sapphire regenerative amplifier (LIBRA-USP-HE) was used for time-resolved Raman measurements. A narrowband picosecond pulses (802 nm, 0.6 nm, 1.2 ps) generated by a home-built grating filter (1200 gr/mm) was used for the Raman pump, and a broadband (850–1000 nm) whitelight continuum generated in a YAG window (Newlight Photonics, Toronto, ON) was used for the Raman probe. The Raman probe filtered with long pass filters (FEL0850, Thorlabs Inc., Newton, NJ; 830 DCLP, Omega Optical Inc., Brattleboro, VT) was combined at the sample with the Raman pump and the actinic pump at 403 nm generated from SHG. The Raman pump was modulated at 500 Hz by an optical chopper (MC2000, Thorlabs Inc.) and the Raman probe was recorded at 1 kHz shot-to-shot level by a spectrograph (Triax 320, Horiba Jobin Yvon GmbH, Bensheim, Germany) and a CCD detector (PIXIS 100, Princeton Instruments, Trento, NJ). The optical time delay between the actinic pump and the Raman pump/probe pair was controlled by a motorized stage (MFN25PP, Newport Inc.) and a controller (ESP300, Newport Inc.). The Raman pump of 350 nJ pulse energy and the actinic pump of 750 nJ pulse energy were used in a typical FSRS measurement.

## 2.5. Computational details

DFT simulations for the Raman vibrational modes of alizarin were conducted by the Gaussian 09 software package (Gaussian Inc., Wallingford, CT), and the B3LYP/6-31G(d,p) level of theory with the optimized geometries from previous TDDFT results was used [23, 38]. The scaling factors for the vibrational frequencies obtained in previous reports were used to visualize the Raman spectra of alizarin both in ground and excited electronic states with arbitrary bandwidths of 10–15 cm$^{-1}$ [39].

# 3. Steady-state absorption and emission spectra of alizarin

In the ground state, the LE tautomer exists lower than the PT tautomer in energy and the energy barrier between two tautomers is too high for the proton transfer in the ground state to be observed [19]. On the other hand, the LE tautomer in the excited state which can be approached by photoexcitation exists higher in energy than the PT tautomer, and the tau-tomerization to the PT tautomer can occur depending on the barrier height separating two tautomers [11–14, 16, 27, 40].

The absorption and emission spectra of alizarin in $n$-heptane, ethanol, and DMSO solution are shown in **Figure 2(a)**. The absorption spectrum of alizarin in $n$-heptane appears as several vibronic bands at ~405, 425, and 450 nm and the absorption bands of alizarin in both ethanol and DMSO are inhomogeneously broadened and red-shifted by 20–30 nm from the bands in $n$-heptane. The emission bands of alizarin in $n$-heptane centered at 610 and 660 nm show large Stokes' shifts from the absorption band representing the intramolecular proton transfer in the excited state. The emission spectra of alizarin in ethanol and DMSO show increased emission in the range of 500–600 nm in addition to main emission bands at 620 and 670 nm, which is interpreted as the emission signal originating from the LE state in the excited state.

**Figure 2(b)** shows the dependence of the excitation wavelength in the emission spectra of alizarin in ethanol. Alizarin shows two emission bands in ethanol solution. One centered at 535 nm from the LE tautomer appears strongly with 485 nm excitation, while the other cen-tered at 620 nm from the PT tautomer becomes the main band with 405-nm excitation. The excess energy in the 405-nm excitation may facilitate the ESIPT by overcoming the energy bar-rier of the LE-PT tautomerization. Concentration and wavelength dependences in the emis-sion spectra of alizarin can be the evidence for the existence of the energy barrier between the LE and PT tautomers [19, 20, 41].

To further investigate the intramolecular proton transfer of alizarin and the solvent depen-dence, the steady-state absorption and emission spectra of alizarin in binary mixtures of etha-nol and water were measured with 405-nm excitation as shown in **Figure 3**. The absorption spectra of alizarin show a slight increase in absorbance with the addition of water to ethanol up to 50% without any spectral change. However, the emission spectra of alizarin show a strong solvent dependence. The PT emission bands at 615 and 670 nm decrease as the fraction of water increases up to 50% while the LE emission band at 530 nm increases. The isosbestic

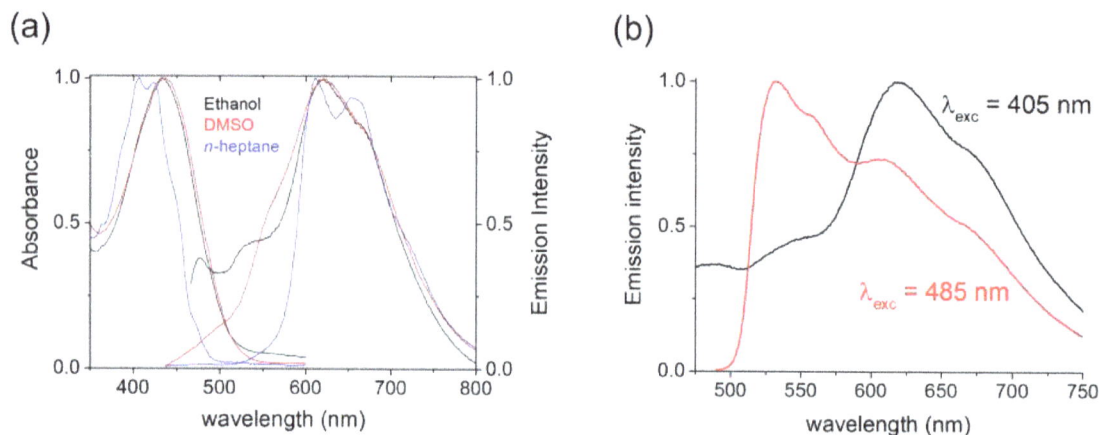

**Figure 2.** (a) Steady-state absorption and emission spectra of alizarin (adapted with permission from ref. [44]. Copyright 2017 ACS), (b) emission spectra of alizarin in ethanol with 405- and 485-nm excitations (adapted with permission from ref. [41]. Copyright 2015 Elsevier).

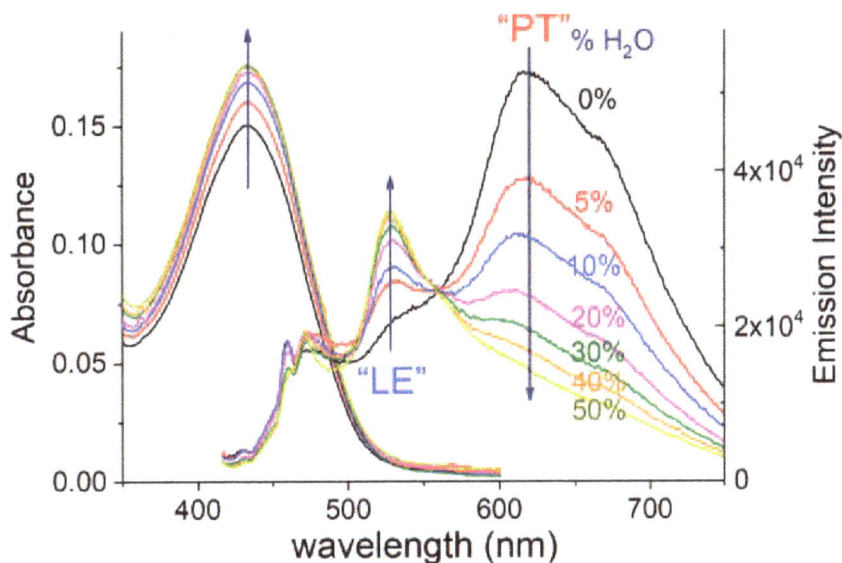

**Figure 3.** Steady-state absorption and emission spectra of alizarin in ethanol and binary mixtures of ethanol and water with 405-nm excitation. Adapted with permission from ref. [41]. Copyright 2015 Elsevier.

point between the LE and PT emission bands is clearly observed at 560 nm, which clearly supports the transition between the LE and PT tautomers. In addition, a decrease of overall quantum yield of alizarin with the addition of water may represent a nonradiative rate constant of the LE tautomer is much smaller than that of the PT tautomer. Recently, the effect of water on the ESIPT reaction of alizarin was further investigated by the simulations based on the time-dependent density functional theory [42]. It has been noted that the strong intramolecular hydrogen bonding of alizarin between the carbonyl and hydroxyl group may facilitate the ESIPT reaction in the excited state. Furthermore, the inhibition of the ESIPT process by water molecules by forming hydrogen bonds with the carbonyl or hydroxyl groups of alizarin was proposed, which weakens the intramolecular hydrogen bonding associated with the ESIPT process and thus increases the energy barrier between the LE and PT tautomer [42].

We have used time-resolved electronic (femtosecond transient absorption) and vibrational spectroscopy (FSRS) to further study the detailed kinetics and mechanism of the ESIPT reaction of alizarin in the excited state.

## 4. Excited state intramolecular proton transfer of alizarin

### 4.1. Femtosecond transient absorption results

Transient absorption results of alizarin in ethanol and in a binary mixture of ethanol:water = 1:1 with 403-nm excitation are shown in **Figure 4**. Within 10 ps time delay, the excited state absorption (ESA) band centered at 510 nm and the stimulated emission (SE) band in the 570–750 nm range are observed in both ethanol and ethanol-water mixture. A broad and weak ESA band in the 500–550 nm range is left after 1 ns time delay for the ethanol-water mixture, while all the excited state population of alizarin in ethanol solution decays to the ground state by the same

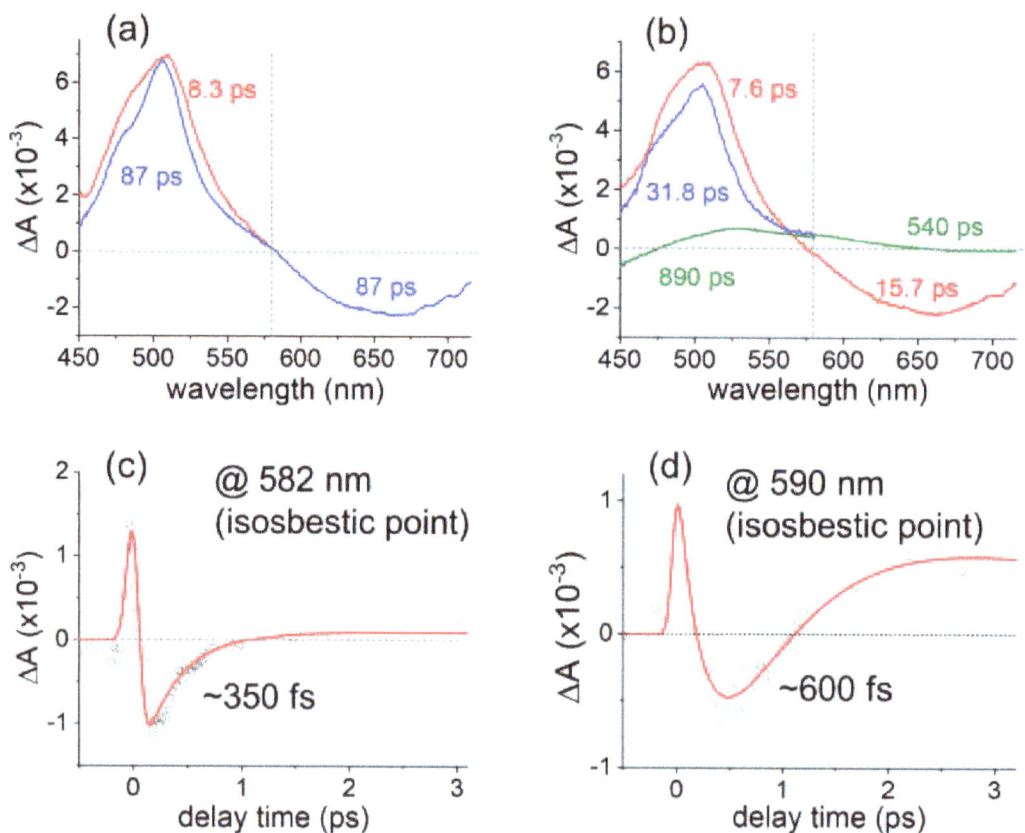

**Figure 4.** Transient absorption results of alizarin (a) in ethanol and (b) in 1:1 mixture of ethanol and water, excited-state kinetics of alizarin (c) in ethanol (582 nm) and (d) in DMSO (590 nm). Adapted from ref. [41]. Copyright 2015 Elsevier.

time. The global fit results for alizarin in ethanol are summarized by two kinetic components of 8.3 and 87 ps whose evolution associated difference spectra (EADS) are shown in **Figure 4(a)**. The 8.3 ps component with slightly broader absorption band (450–580 nm) but without emission signal represents the vibrationally hot PT tautomer, and the 87 ps component with both absorption and emission (580–750 nm) signals represents the relaxed PT tautomer in the excited state, which is consistent with previous results [20, 24–26].

The excited state dynamics of alizarin in ethanol-water mixture is somewhat complex. Instead of performing the global fit analysis of the whole transient absorption data, we analyzed the absorption (<580 nm) and emission (>580 nm) part of the data separately in the global fit analysis. We found three kinetic components of 7.6, 31.8, and 890 ps from the absorption part and two components of 15.7 and 540 ps from the emission part of the data. The kinetic components of 7.6 and 31.8 ps in the absorption part are tentatively assigned as the vibrationally hot and relaxed PT tautomers of alizarin, respectively, by inferring from the results of ethanol solution. However, the 7.6 ps component may include the decay of the LE state, as the blocking of the proton transfer reaction was observed with the addition of water from results of the steady-state emission spectra. The 15.7 ps lifetime of the first emission component of the data is much shorter than the lifetime of 31.8 ps component in the absorption part, thus this component may also represent the emission signal of both the LE and PT tautomer which cannot be separated in all the analysis we have done. In addition to the fast kinetic components for the LE and PT tautomers, a long-lived component (540 or 890 ps) appeared as a very broad absorption band in 500–600 nm.

It is noted that the shortened lifetime of the PT tautomer (87 → 31.8 ps) with the addition of water to ethanol observed in transient absorption measurements is consistent to the reduced quantum yield and the increased nonradiative rate constant of alizarin observed in the steady-state emission measurements. As suggested by the recent theoretical study [42], water molecules may form hydrogen bonds with the carbonyl and hydroxyl groups of alizarin and impede the intramolecular proton transfer reaction of alizarin. Thus the long decay component in the transient absorption of alizarin in ethanol-water mixture may be considered as the "trapped" state of alizarin with water molecules. Further details on the solute-solvent interaction and resulting ESIPT kinetics can be investigated by FSRS, where time-resolved structural changes of solute and solvent molecules can be monitored.

It has been proposed that faster components of 300–400 fs time constant generally observed from the transient absorption signals of alizarin in the wavelengths (570–585 nm) where the strong ESA and SE signals cancel out, may represent the kinetics for the vibrational relaxation in the LE tautomer and the ESIPT to the PT tautomer [41, 43]. We also observed these fast components universally in the transient absorption results of alizarin in ethanol, methanol, DMSO, and ethanol-water mixture (examples are shown in **Figure 4(c)** and **(d)**), but did not show any dependence on the solvent polarity. Since the ESA, SE, and the ground-state bleaching signals of two tautomers of alizarin in transient absorption measurements are overlapped in wavelength and time, it seems to be very difficult to separate the kinetic components of the vibrational relaxation of the LE and PT tautomers, the ESIPT, etc. We conclude that the transient absorption measurements may be inadequate for the correct analysis of the ESIPT process, a further investigation by FSRS was performed to obtain the population and structural dynamics of alizarin upon photoexcitation.

## 4.2. Femtosecond stimulated Raman results

### 4.2.1. FSRS details

Alizarin is soluble in ethanol and DMSO but the Raman spectrum of ethanol overlaps that of alizarin in many spectral regions. Then small changes in the vibrational modes of alizarin might not be observed in ethanol solution due to strong Raman modes of ethanol. The Raman bands of DMSO, however, can be separable from the Raman modes of alizarin. Thus femtosecond stimulated Raman measurements of alizarin were done with DMSO solution. From the analysis of transient absorption result of alizarin in DMSO solution, two kinetic components were obtained [44]. Two components of 1.1 and 83.3 ps represent the vibrational relaxation in the PT tautomer and the lifetime of PT tautomer in the excited state, respectively. Although a fast (~600 fs) kinetic component was observed at 590 nm where the strong ESA and SE signals cancel out, it is not clear whether this component represents the ESIPT dynamics of alizarin.

The Raman intensity of the FSRS, often called the Raman gain can be evaluated by Eq. (1):

$$(\text{Raman Gain}) = \frac{I_{\text{R.Pump-ON}} - I_{\text{Bkg}}}{I_{\text{R.Pump-OFF}} - I_{\text{Bkg}}} \tag{1}$$

where $I_{\text{R.Pump-ON}}$ and $I_{\text{R.Pump-OFF}}$ represent the intensity of Raman probe with and without the Raman pump, respectively, and $I_{\text{Bkg}}$ represents the dark signal of the CCD detector. The Raman probe of 600,000 pulses (60 accumulations of 10-sec acquisition) was averaged in a typical Raman gain measurement with the half of them focused together with the Raman pump to the sample, to obtain a Raman gain signal in a signal-to-noise level of $2 \times 10^{-5}$ (or 0.002%) at a specific time delay with the actinic pump pulses. Time-resolved stimulated Raman spectra of alizarin in DMSO at multiple time delays were obtained at time delays of −1 to 100 ps and the ground state spectrum measured at −10 ps time delay, for example, was subtracted from each stimulated Raman spectrum to obtain the difference stimulated Raman spectra shown in **Figure 5(a)**. A small portion of transient absorption signal, for example, the ESA and SE can be obtained together with stimulated Raman gain signals at most time delays, thus a polynomial background subtraction was performed to remove the transient absorption signal.

### 4.2.2. The population and structural dynamics of the ESIPT

Major Raman bands of the ground electronic state of alizarin in the 1500–1800 cm$^{-1}$ range shown in **Figure 5(a)** were assigned as the ring $\nu$(C=C) at 1573 and 1594 cm$^{-1}$, and $\nu$(C=O) at 1634 and 1661 cm$^{-1}$ mainly according to the DFT simulation results. One $\nu$(C=O) at 1661 cm$^{-1}$ is assigned to the isolated carbonyl at C10 position and the other at 1634 cm$^{-1}$ is the carbonyl at the site of the ESIPT and adjacent to a hydroxyl group [44]. Another Raman band at 1191 cm$^{-1}$ is assigned as $\delta$(CH) and $\delta$(OH). In the excited stimulated Raman spectra of alizarin, several Raman bands at 1162, 1555, and 1632 cm$^{-1}$ appeared in 50–100 fs after the photoexcitation and showed a decay after 20 ps or so. According to the TDDFT simulation results [23, 44], we tentatively assign the 1162 cm$^{-1}$ band as the $\delta$(CH) and $\delta$(OH) of the PT tautomer, and the 1555 and 1632 cm$^{-1}$ as $\nu$(C=C) and $\nu$(C=O) bands also in the PT tautomer of alizarin.

To obtain the details of the excited state dynamics and the ESIPT from the stimulated Raman bands of alizarin, the experimental data were fit with a low-order polynomial background and several Gaussian functions for Raman bands. The population dynamics of $\nu$(C=C) and $\nu$(C=O) bands at 1555 and 1632 cm$^{-1}$ shown in **Figure 5(b)** shows a ubiquitous sharp rise in 70–80 fs and a slow decay into the ground state which is compatible to the PT tautomer's lifetime of 83.3 ps as shown in the transient absorption results. The structural dynamics of $\nu$(C=C) and $\nu$(C=O) modes of the PT tautomer are shown in **Figure 5(c)** and **(d)** as the time-dependent changes in the peak position and the bandwidth. The peak shift of the solvent vibrational mode, $\delta$(CH$_3$) of DMSO at 1421 cm$^{-1}$ represents the instrument response function of FSRS measurements for comparison with the excited dynamics of alizarin Raman bands. It is interesting to note that a strong blue-shift (1540 → 1553 cm$^{-1}$) and a decrease of bandwidth (28 → 24 cm$^{-1}$) of $\nu$(C=C) band all occur in an ultrafast time scale of ~150 fs after photoexcitation. On the other hand, the $\nu$(C=O) band shows a strong red-shift (1645 → 1630 cm$^{-1}$) in the same time delay of 100–150 fs although this vibrational band appears too broad for the bandwidth analysis. As well represented in **Figure 5(b–d)**, the population growth and the structural changes of $\nu$(C=C) and $\nu$(C=O) Raman bands of the PT tautomer of alizarin are interpreted as the ESIPT process from the LE to the PT tautomer. Since no Raman band of the LE tautomer has been identified from the FSRS results, this could also be included for the dynamics of $\nu$(C=C) and $\nu$(C=O) bands at 1555 and 1632 cm$^{-1}$. The vibrational relaxation along an electronic potential surface would generally result in slight blue-shifts in the strongly coupled vibrational modes due to the anharmonicity of the electric potential surface. However, the $\nu$(C=C) and $\nu$(C=O) bands showed strong (~15 cm$^{-1}$) peak shifts either in increasing and decreasing bandwidth, respectively, during the ultrafast period of the population

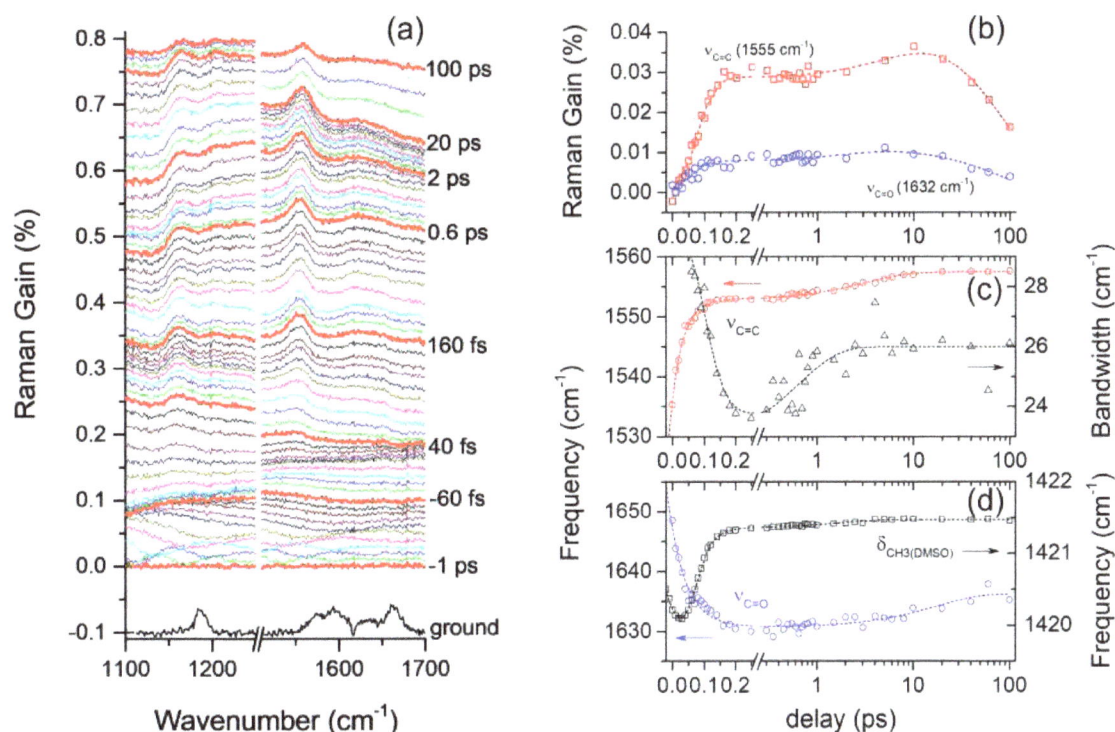

**Figure 5.** (a) Time-resolved stimulated Raman spectra of alizarin in DMSO following 403-nm photoexcitation, (b) population dynamics for the Raman bands of $\nu$(C=C) at 1555 cm$^{-1}$ and $\nu$(C=O) at 1632 cm$^{-1}$, (c) $\nu$(C=C) and (d) $\nu$(C=O) Raman bands from the PT conformer of alizarin in the excited state and peak shift of a solvent $\delta$(CH$_3$) mode which represents the instrument response function of FSRS measurements. Adapted with permission from ref. [44]. Copyright 2017 ACS.

growth for the PT tautomer. This cannot be explained by any type of relaxation inside the same potential surface but has to be understood as the nuclear rearrangements for the intramolecular proton transfer reaction. Therefore, we conclude that the intramolecular proton transfer reaction of alizarin in the excited state occurs in ultrafast time scale of 70–80 fs.

Another interesting fact is the strong and opposite peak shifts observed for $\nu$(C=C) and $\nu$(C=O) bands during the ESIPT reaction. We could imagine a transition state for the ESIPT reaction of alizarin as a six-membered ring formed by intramolecular hydrogen bonding between carbonyl group and hydroxyl group. We propose that the strong and opposite peak shifts of the $\nu$(C=C) and $\nu$(C=O) band directly represent the changes in the resonance structure of the alizarin backbone which is composed of multiple C=C and C-C bonds and a C=O. The details of the ESIPT reaction mechanism of alizarin need be confirmed by thorough theoretical investigations, which is beyond of the scope of this chapter. The reaction mechanism of many ESIPT reactions and the existence of transition states have been recently reported by several theoretical works based on TDDFT and several transition states of six-membered ring between carbonyl and hydroxyl groups were represented for 1,8-dihydroxy-2-naphthaldehyde [7, 45, 46]. Although a separate transition state of the ESIPT reaction was not resolved from the FSRS results, we also propose the reaction may occur via the transition state of a new hydrogen-bond six-membered ring attached to the anthraquinone backbone.

As shown in **Figure 5(b–d)**, two more kinetic components other than the population decay of the PT tautomer were identified. A slight blue-shift (1553 → 1557 cm$^{-1}$) and an increased bandwidth (24 → 26 cm$^{-1}$) of $\nu$(C=C) mode observed in 3–10 ps and another slight blue-shift

(1630 → 1636 cm$^{-1}$) of ν(C=O) mode shown in 20–30 ps represent the vibrational relaxation in the product potential surface of the PT conformer. There were no further changes in peak position and bandwidth of ν(C=C) and ν(C=O) modes during the population decay of the PT tautomer, which also supports the assignment of the vibrational relaxation in the PT potential surface.

## 5. Solvation dynamics

The intramolecular proton transfer reaction of alizarin in the excited state was evidenced by the population and structural dynamics of two major vibrational modes of ν(C=C) and ν(C=O). Ultrafast ESIPT reaction of alizarin in the excited state can also be observed indirectly by the changes in the solvent vibrational spectrum such as the instantaneous disruption of the solvation shells and the formation of new solvation. **Figure 6(a)** and **(b)** show the difference stimulated Raman spectra of the solvent DMSO, especially ν(S=O) at 1044 cm$^{-1}$ with alizarin concentrations of 20 and 1 mM, respectively. Solvent DMSO is known to form hydrogen bonds in solution between S=O and C-H groups, and also a polymeric structure is formed at low temperature [47]. The Raman band of ν(S=O) is composed of multiple subbands including the symmetric (1026 cm$^{-1}$) and antisymmetric stretching (1042 cm$^{-1}$) of dimer, the symmetric stretching (1058 cm$^{-1}$) of monomer, etc., [47, 48]. In the DMSO solutions of alizarin in 0–15 mM concentration range, the ν(S=O) band of DMSO shows strong peak shifts (1042 → 1024 cm$^{-1}$), which represents changes in the hydrogen bonding network of DMSO [44]. The δ(CH$_3$) band of DMSO shows no major spectral changes upon the alizarin concentration, thus the solvation of alizarin with DMSO mainly occurs via hydrogen bonds with the sulfoxide group of DMSO [44].

As shown in **Figure 6(a)** and **(b)**, a sharp dispersive pattern in the ν(S=O) band appears instantly with the actinic pulse for both 20 and 1 mM concentrations of alizarin. The Raman intensity for the symmetric stretching of monomer around 1060 cm$^{-1}$ decreases and the symmetric stretching of dimer around 1025 cm$^{-1}$ increases, which clearly shows the instantaneous changes in hydrogen bonding of DMSO molecules. As clearly seen with the 1-mM solution case, this dispersive pattern disappears very quickly as the actinic pulse leaves the solution. In other words, a disruption in the hydrogen bonding of DMSO molecules created by ultrafast laser pulses is removed quickly by reforming hydrogen bonds between DMSO molecules. It seems that the hydrogen bond reformation occurs much faster than the instrument response function of FSRS (~100 fs). Considering from the ultrafast dynamics of the dispersive signals of ν(S=O) band, the nonpolar solvation effect may be understood as the origin of this sharp dispersive signal [49–51]. We have also observed a similar dispersive background signal in the δ(CH$_3$) band of DMSO [44].

On the other hand, the 20-mM alizarin results showed clearly distinct dynamics for the ν(S=O) band as shown in **Figure 6(a)**. The initial dispersive Raman signals were almost removed in about 100 fs then another type of dispersive Raman signals appeared, which is composed of a small bleaching with almost the same spectral shape as the ground state ν(S=O) band and a much broader positive signal around 990 cm$^{-1}$. **Figure 7** clearly shows this dispersive Raman pattern in the ν(S=O) of DMSO appearing 100 fs after the actinic pump, where the initial dispersive Raman signals of the ν(S=O) obtained with 1-mM alizarin solution were subtracted from the results with 20-mM alizarin solution. The bleaching of the ground state ν(S=O) Raman band may result from the local heating due to the vibrational cooling of solute molecules, then the recovery of the

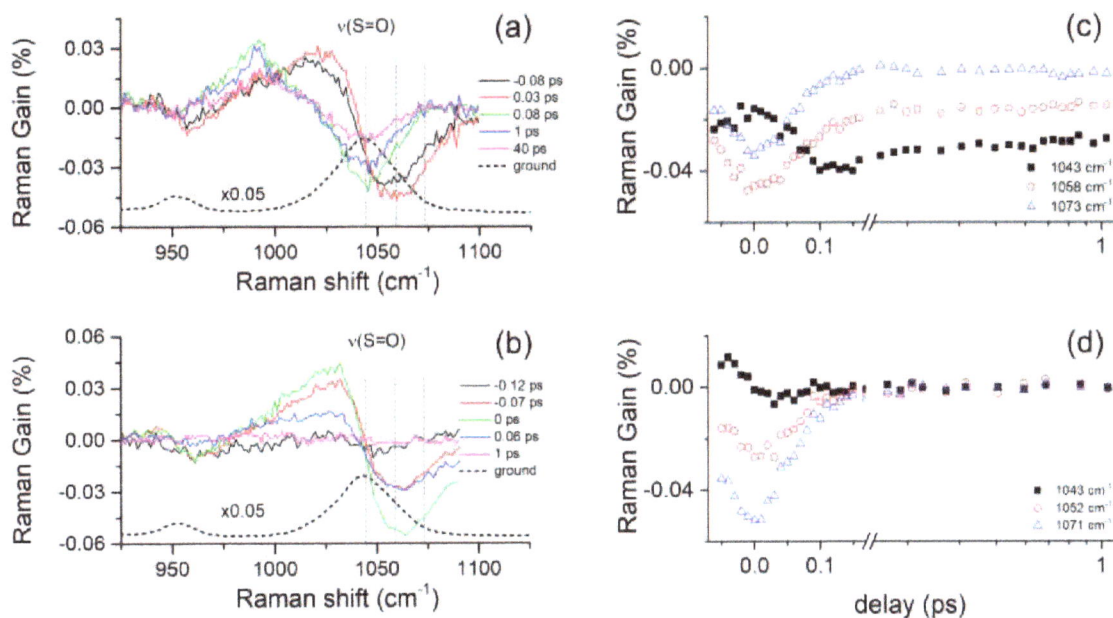

**Figure 6.** Difference stimulated Raman spectra of ν(S=O) mode of DMSO in (a) 20- and (b) 1-mM alizarin solution, the excited-state kinetics of ν(S=O) bands at several frequencies in (c) 20- and (d) 1-mM alizarin solution. Adapted from ref. [44]. Copyright 2017 ACS.

bleaching signals by the local cooling would take several tens of picoseconds [51–53]. The decay of the second dispersive Raman signals at 990 and 1043 cm$^{-1}$ is compatible to the local cooling time but a huge frequency difference (~50 cm$^{-1}$) cannot be explained by the local heating of solvent DMSO. In a control experiment to measure temperature-dependent Raman spectra for the ν(S=O) of DMSO, the spectral changes of less than 5 cm$^{-1}$ were observed with temperature increase of 40–50 °C. Therefore, we conclude that the second dispersive Raman signals of ν(S=O) appearing at 100 fs time delay and between 990 and 1043 cm$^{-1}$ cannot be explained as the local heating of solvent molecules pumped by vibrationally cooled solute molecules and the changes in the solvation

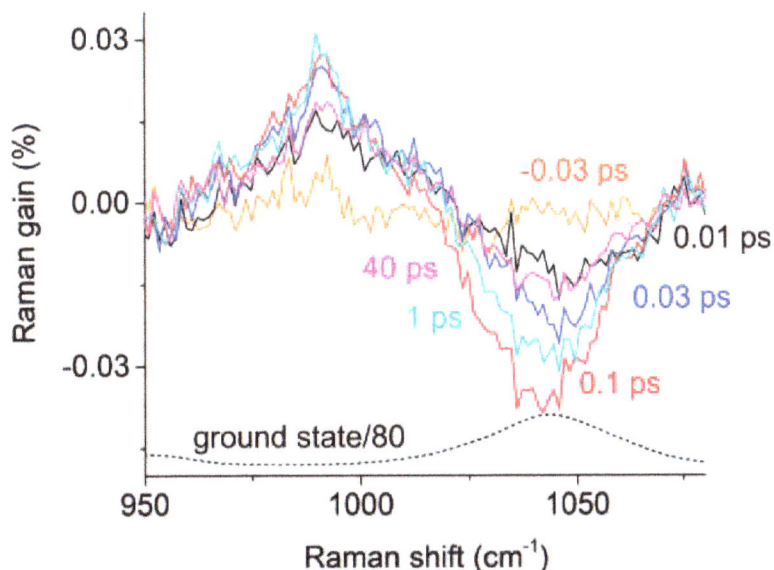

**Figure 7.** Dispersive Raman signals of ν(S=O) of DMSO. The initial dispersive pattern due to nonpolar solvation changes of DMSO molecules is subtracted.

shells of DMSO molecules due to the ESIPT reaction of alizarin molecules from the LE to the PT tautomers must be considered. Different from the initial dispersive signals in the $\nu$(S=O) and $\delta$(CH$_3$) bands of DMSO, the second dispersive signals in the $\nu$(S=O) must be considered as originating from the polar or dielectric type of solvation [51]. The dispersive Raman signal of the $\nu$(S=O) band at 1044 cm$^{-1}$ showed a growth with a $60 \pm 30$ fs time constant and a decay with a $4.9 \pm 1.5$ ps time constant, which clearly shows that the hydrogen-bonding network of DMSO was created by the ESIPT reaction of alizarin and decayed as the vibrational relaxation of the product along the potential surface of the PT tautomer in the excited state. In this work, we showed that the $\nu$(S=O) band of solvent DMSO can be used to determine the ultrafast ESIPT reaction and the subsequent vibrational relaxation in the reaction product. The breaking and reforming of hydrogen-bonding network of DMSO can be successfully observed by the $\nu$(S=O) band of DMSO thus this method can also be applied to many chemical reactions occurring in the photoinduced excited states.

## 6. Conclusion

In this chapter, the ESIPT reaction and the excited state dynamics of alizarin were explored by time-resolved electronic and vibrational spectroscopy with the femtosecond time-resolution. The dependence on solvent polarity and excitation wavelength was observed in the steady-state emission spectra of alizarin, where the barrier height between the LE and PT tautomers in the excited state may exist and be controlled by the solvent polarity. The transient absorption results of alizarin in ethanol and ethanol-water mixture were so complicated and overlapping, so the ESIPT rate constant from the LE to the PT tautomers was not separable from the vibrational relaxation and population decay of both tautomers. Instead, the ESIPT of alizarin in the excited state was clearly observed in femtosecond stimulated Raman measurements. The population and structural dynamics of two major vibrational modes of $\nu$(C=C) and $\nu$(C=O) clearly showed the dynamics of the ESIPT rate to the PT tautomer, the vibrational relaxation and the population decay of the product PT tautomer. The vibrational signature of the LE tautomer was not observed in FSRS, but the reaction mechanism of the ESIPT including a transition state of a newly formed six-membered ring composed of the carbonyl and hydroxyl groups was estimated by the strong and opposite peak shifts of $\nu$(C=C) and $\nu$(C=O) seen in the stimulated Raman spectra of alizarin during the reaction. From the population growth and structural transformation into the PT tautomer, we concluded that the ESIPT of alizarin occurs in an ultrafast time scale of 70–80 fs. During the ESIPT reaction of alizarin, solvent DMSO molecules showed ultrafast structural changes involving hydrogen bonds with solute molecules. When the solute concentration is very low, DMSO shows a dispersive Raman signal in the $\nu$(S=O) and $\delta$(CH$_3$) modes only with the actinic pump. The instantaneous disruption and reformation of hydrogen bonds may suggest a nonpolar type of solvation between solvent molecules. On the other hand, complicated dispersive Raman signals in the $\nu$(S=O) mode of DMSO were observed with a concentrated (20 mM) solution of alizarin. After the same instantaneous solvent responses completed in 100 fs, the second dispersive Raman pattern with a bleaching of the ground state spectrum appeared and decayed with 60 fs and 5 ps time scales. This also represents the disruption of hydrogen bonds of DMSO molecules, more specifically between solute molecules and in the polar or dielectric solvation shells. Interestingly, the dynamics for the ultrafast proton transfer reaction and the vibrational relaxation in the product state was measured by the solvation signals of solvent DMSO.

# Acknowledgements

This work was supported by Basic Science Research Program funded by the Ministry of Education (2017R1A1D1B03027870, 2014R1A1A2058409) and by the International Cooperation Program (2016K2A9A1A01951845), through the National Research Foundation of Korea (NRF). The GIST Research Institute (GRI) in 2018 and the PLSI supercomputing resources of the Korea Institute of Science and Technology Information also supported this research.

# Conflict of interest

The authors declare no conflict of interest.

# Author details

Sebok Lee[1†], Myungsam Jen[1†], Kooknam Jeon[1], Jaebeom Lee[2], Joonwoo Kim[1] and Yoonsoo Pang[1*]

*Address all correspondence to: ypang@gist.ac.kr

1 Department of Chemistry, Gwangju Institute of Science and Technology, Gwangju, Republic of Korea

2 Department of Physics and Photon Science, Gwangju Institute of Science and Technology, Gwangju, Republic of Korea

†These authors contributed equally

# References

[1] Rini M, Magnes BZ, Pines E, Nibbering ETJ. Real-time observation of bimodal proton transfer in acid-base pairs in water. Science. 2003;**301**:349-352. DOI: 10.1126/science.1085762

[2] Siwick BJ, Cox MJ, Bakker HJ. Long-range proton transfer in aqueous acid–base reactions. Journal of Physical Chemistry B. 2008;**112**:378-389. DOI: 10.1021/jp075663i

[3] Perez-Lustres JL, Rodriguez-Prieto F, Mosquera M, Senyushkina TA, Ernsting NP, Kovalenko SA. Ultrafast proton transfer to solvent: Molecularity and intermediates from solvation- and diffusion-controlled regimes. Journal of the American Chemical Society. 2007;**129**:5408-5418. DOI: 10.1021/ja0664990

[4] Elsaesser T, Kaiser W, Luettke W, Luttke W. Picosecond spectroscopy of intramolecular hydrogen bonds in 4,4',7,7'-Tetramethyllndigo. Journal of Physical Chemistry. 1986;**90**:2901-2905. DOI: 10.1021/j100404a024

[5] Ernsting NP. Dual fluorescence and exclted-state intramolecular proton transfer in jet-cooled 2,5-Bis(2-benzoxazolyl) hydroquinone. Journal of Physical Chemistry. 1985; **89**:4932-4939. DOI: 10.1021/j100269a010

[6] Simkovitch R, Huppert D. Excited-state intramolecular proton transfer of the natural product quercetin. Journal of Physical Chemistry B. 2015;**119**:10244-10251. DOI: 10.1021/acs.jpcb.5b04867

[7] Tseng HW, Liu JQ, Chen YA, Chao CM, Liu KM, Chen CL, et al. Harnessing excited-state intramolecular proton-transfer reaction via a series of amino-type hydrogen-bonding molecules. Journal of Physical Chemistry Letters. 2015;**6**:1477-1486. DOI: 10.1021/acs.jpclett.5b00423

[8] Takeuchi S, Tahara T. Coherent nuclear wavepacket motions in ultrafast excited-state intramolecular proton transfer: Sub-30-fs resolved pump-probe absorption spectroscopy of 10-hydroxybenzo[h]quinoline in solution. Journal of Physical Chemistry A. 2005;**109**:10199-10207. DOI: 10.1021/jp0519013

[9] Kim CH, Joo T. Coherent excited state intramolecular proton transfer probed by time-resolved fluorescence. Physical Chemistry Chemical Physics. 2009;**11**:10266-10269. DOI: 10.1039/b915768a

[10] Lochbrunner S, Wurzer AJ, Riedle E. Microscopic mechanism of ultrafast excited-state intramolecular proton transfer: A 30-fs study of 2-(2′-hydroxyphenyl)benzothiazole. Journal of Physical Chemistry A. 2003;**107**:10580-10590. DOI: 10.1021/jp035203z

[11] Flom SR, Barbara PF. Proton transfer and hydrogen bonding in the internal conversion of S1 Anthraquinones. Journal of Physical Chemistry. 1985;**89**:4489-4494. DOI: 10.1021/j100267a017

[12] Miliani C, Romani A, Favaro G. Acidichromic effects in 1,2-di- and 1,2,4-trihydroxyanthraquinones. A spectrophotometric and fluorimetric study. Journal of Physical Organic Chemistry. 2000;**13**:141-150. DOI: 10.1002/(SICI)1099-1395(200003)13:3

[13] Mech J, Grela M, Szaciłowski K. Ground and excited state properties of alizarin and its isomers. Dyes and Pigments. 2014;**103**:202-213. DOI: 10.1016/j.dyepig.2013.12.009

[14] Le Person A, Cornard JP, Say-Liang-Fat S. Studies of the tautomeric forms of alizarin in the ground state by electronic spectroscopy combined with quantum chemical calculations. Chemical Physics Letters. 2011;**517**:41-45. DOI: 10.1016/j.cplett.2011.10.015

[15] Neuwahl FVR, Bussotti L, Righini R, Buntinx G. Ultrafast proton transfer in the S1 state of 1-chloroacetylaminoanthraquinone. Physical Chemistry Chemical Physics. 2001;**3**:1277-1283. DOI: 10.1039/b007312l

[16] Cho SH, Huh H, Kim HM, Kim NJ, Kim SK. Infrared-visible and visible-visible double resonance spectroscopy of 1-hydroxy-9,10-anthraquinone-(H2O)n (n=1,2) complexes. Journal of Chemical Physics. 2005;**122**:34305. DOI: 10.1063/1.1829991

[17] Habeeb MM, Alghanmi RM. Spectrophotometric study of intermolecular hydrogen bonds and proton transfer complexes between 1,2-dihydroxyanthraquinone and some

aliphatic amines in methanol and acetonitrile. Journal of Chemical & Engineering Data. 2010;**55**:930-936. DOI: 10.1021/je900528h

[18] Cysewski P, Jeliński T, Przybyłek M, Shyichuk A. Color prediction from first principle quantum chemistry computations: A case of alizarin dissolved in methanol. New Journal of Chemistry. 2012;**36**:1836-1843. DOI: 10.1039/c2nj40327g

[19] Sasirekha V, Umadevi M, Ramakrishnan V. Solvatochromic study of 1,2-dihydroxyanthraquinone in neat and binary solvent mixtures. Spectrochimica Acta - Part A: Molecular and Biomolecular Spectroscopy. 2008;**69**:148-155. DOI: 10.1016/j.saa.2007.03.021

[20] Reta MR, Anunziata JD, Cattana RI, Silber JJ. Comparison between solvatochromic and chromatographic studies of anthraquinones in binary aqueous mixtures. Analytica Chimica Acta. 1995;**306**:81-89. DOI: 10.1016/0003-2670(94)00595-D

[21] Zhao J, Ji S, Chen Y, Guo H, Yang P. Excited state intramolecular proton transfer (ESIPT): From principal photophysics to the development of new chromophores and applications in fluorescent molecular probes and luminescent materials. Physical Chemistry Chemical Physics. 2012;**14**:8803-8817. DOI: 10.1039/C2CP23144A

[22] Kwon JE, Park SY. Advanced organic optoelectronic materials: Harnessing excited-state intramolecular proton transfer (ESIPT) process. Advanced Materials. 2011;**23**:3615-3642. DOI: 10.1002/adma.201102046

[23] Amat A, Miliani C, Romani A, Fantacci S. DFT/TDDFT investigation on the UV-vis absorption and fluorescence properties of alizarin dye. Physical Chemistry Chemical Physics. 2015;**17**:6374-6382. DOI: 10.1039/c4cp04728a

[24] Dworak L, Matylitsky VV, Wachtveitl J. Ultrafast photoinduced processes in alizarin-sensitized metal oxide mesoporous films. ChemPhysChem. 2009;**10**:384-391. DOI: 10.1002/cphc.200800533

[25] Matylitsky VV, Lenz MO, Wachtveitl J. Observation of pH-dependent back-electron-transfer dynamics in alizarin/TiO2 adsorbates: Importance of trap states. Journal of Physical Chemistry B. 2006;**110**:8372-8379. DOI: 10.1021/jp060588h

[26] Huber R, Moser J, Gratzel M, Wachtveitl J. Real-time observation of photoinduced adiabatic Electron transfer in strongly coupled dye/semiconductor colloidal systems with a 6 fs time constant. Journal of Physical Chemistry B. 2002;**106**:6494-6499. DOI: 10.1021/jp0155819

[27] Choi JR, Jeoung SC, Cho DW. Two-photon-induced excited-state intramolecular proton transfer process in 1-hydroxyanthraquinone. Chemical Physics Letters. 2004;**385**:384-388. DOI: 10.1016/j.cplett.2004.01.011

[28] Ryu J, Woo Kim H, Soo Kim M, Joo T. Ultrafast excited state intramolecular proton transfer dynamics of 1-hydroxyanthraquinone in solution. Bulletin of the Korean Chemical Society. 2013;**34**:465-469. DOI: 10.5012/bkcs.2013.34.2.465

[29] Zhu L, Liu W, Fang C. A versatile femtosecond stimulated Raman spectroscopy setup

with tunable pulses in the visible to near infrared. Applied Physics Letters. 2014;**105**:41106. DOI: 10.1063/1.4891766

[30] Kovalenko SA, Dobryakov AL, Ernsting NP. An efficient setup for femtosecond stimulated Raman spectroscopy. Review of Scientific Instruments. 2011;**82**:63102. DOI: 10.1063/1.3596453

[31] Kukura P, McCamant DW, Mathies RA. Femtosecond stimulated Raman spectroscopy. Annual Review of Physical Chemistry. 2007;**58**:461-488. DOI: 10.1146/annurev. physchem.58.032806.104456

[32] McAnally MO, McMahon JM, Van Duyne RP, Schatz GC. Coupled wave equations theory of surface-enhanced femtosecond stimulated Raman scattering. Journal of Chemical Physics. 2016;**145**:94106. DOI: 10.1063/1.4961749

[33] Hoffman DP, Mathies RA. Femtosecond stimulated Raman exposes the role of vibrational coherence in condensed-phase photoreactivity. Accounts of Chemical Research. 2016;**49**:616-625. DOI: 10.1021/acs.accounts.5b00508

[34] Frontiera RR, Mathies RA. Femtosecond stimulated Raman spectroscopy. Laser & Photonics Reviews. 2011;**5**:102-113. DOI: 10.1002/lpor.200900048

[35] Lee I, Lee S, Pang Y. Excited-state dynamics of carotenoids studied by femtosecond transient absorption spectroscopy. Bulletin of the Korean Chemical Society. 2014;**35**:851-857. DOI: 10.5012/bkcs.2014.35.3.851

[36] Jen M, Lee S, Pang Y. Excited-state dynamics of all-trans-retinal investigated by time-resolved electronic and vibrational spectroscopy. Bulletin of the Korean Chemical Society. 2015;**36**:900-905. DOI: 10.1002/bkcs.10168

[37] Van Stokkum IHM, Larsen DS, Van Grondelle R. Global and target analysis of time-resolved spectra. Biochimica et Biophysica Acta, Bioenergetics. 2004;**1657**:82-104. DOI: 10.1016/j.bbabio.2004.04.011

[38] Frisch MJ, Trucks GW, Schlegel HB, Scuseria GE, Robb MA, Cheeseman JR, et al. Gaussian 09, Revision A.02. Wallingford CT: Gaussian, Inc; 2009

[39] Irikura KK, Johnson RD, Kacker RN. Uncertainties in scaling factors for ab initio vibrational frequencies. Journal of Physical Chemistry A. 2005;**109**:8430-8437. DOI: 10.1021/ jp052793n

[40] Smith TP, Zaklika KA, Thakur K, Walker GC, Tominaga K, Barbara PF. Spectroscopic studies of excited-state intramolecular proton-transfer in 1-(acylamino)anthraquinones. Journal of Physical Chemistry. 1991;**95**:10465-10475. DOI: 10.1021/j100178a038

[41] Lee S, Lee J, Pang Y. Excited state intramolecular proton transfer of 1,2-dihydroxyanthraquinone by femtosecond transient absorption spectroscopy. Current Applied Physics. 2015;**15**:1492-1499. DOI: 10.1016/j.cap.2015.08.017

[42] Peng Y, Ye Y, Xiu X, Sun S. Mechanism of excited-state intramolecular proton transfer for 1,2-dihydroxyanthraquinone: Effect of water on the ESIPT. Journal of Physical Chemistry A. 2017;**121**:5625-5634. DOI: 10.1021/acs.jpca.7b03877

[43] Huber R, Spörlein S, Moser JE, Grätzel M, Wachtveitl J. The role of surface states in the ultrafast photoinduced electron transfer from sensitizing dye molecules to semiconductor colloids. Journal of Physical Chemistry B. 2000;**104**:8995-9003. DOI: 10.1021/jp9944381

[44] Jen M, Lee S, Jeon K, Hussain S, Pang Y. Ultrafast intramolecular proton transfer of alizarin investigated by femtosecond stimulated Raman spectroscopy. Journal of Physical Chemistry B. 2017;**121**:4129-4136. DOI: 10.1021/jp0155819

[45] Peng CY, Shen JY, Chen YT, Wu PJ, Hung WY, Hu WP, et al. Optically triggered stepwise double-proton transfer in an intramolecular proton relay: A case study of 1,8-Dihydroxy-2-naphthaldehyde. Journal of the American Chemical Society. 2015;**137**:14349-14357. DOI: 10.1021/jacs.5b08562

[46] Liu X, Zhao J, Zheng Y. Insight into the excited-state double proton transfer mechanisms of doxorubicin in acetonitrile solvent. RSC Advances. 2017;**7**:51318-51323. DOI: 10.1039/C7RA08945G

[47] Martens WN, Frost RL, Kristof J, Kloprogge JT. Raman spectroscopy of dimethyl sulphoxide and deuterated dimethyl sulphoxide at 298 and 77 K. Journal of Raman Spectroscopy. 2002;**33**:84-91. DOI: 10.1002/jrs.827

[48] Singh S, Krueger PJ. Raman spectral studies of aqueous solutions of non-electrolytes: Dimethylsulfoxide, acetone and acetonitrile. Journal of Raman Spectroscopy. 1982;**13**:178-188. DOI: 10.1002/jrs.1250130214

[49] Berg M. Viscoelastic continuum model of nonpolar solvation. 1. Implications for multiple time scales in liquid dynamics. Journal of Physical Chemistry A. 1998;**102**:17-30. DOI: 10.1021/jp9722061

[50] Stephens MD, Saven JG, Skinner JL. Molecular theory of electronic spectroscopy in nonpolar fluids: Ultrafast solvation dynamics and absorption and emission line shapes. Journal of Chemical Physics. 1997;**106**:2129-2144. DOI: 10.1063/1.473144

[51] Rosspeintner A, Lang B, Vauthey E. Ultrafast photochemistry in liquids. Annual Review of Physical Chemistry. 2013;**64**:247-271. DOI: 10.1146/annurev-physchem-040412-110146

[52] Deàk JC, Pang Y, Sechler TD, Wang Z, Dlott DD. Vibrational energy transfer across a reverse micelle surfactant layer. Science. 2004;**306**:473-476. DOI: 10.1126/science.1102074

[53] Dlott DD. Vibrational energy redistribution in polyatomic liquids: 3D infrared-Raman spectroscopy. Chemical Physics. 2001;**266**:149-166. DOI: 10.1016/S0301-0104(01)00225-7

# 8

# Kinetics and Mechanism of Photoconversion of N-Substituted Amides of Salicylic Acid

Nadezhda Mikhailovna Storozhok and
Nadezhda Medyanik

## Abstract

Studied using optical spectroscopy, stationary, and nanosecond laser photolysis (Nd:YAG laser 355 nm) conversion products in heptane of N-substituted amides of salicylic acid: N-(4-hydroxyhydro-3,5-di-tert-butylphenyl) amide 2-hydroxy-3-t-butyl-5-ethylbenzoic acid (1), 2-(4-hydroxy-3,5-hydroxybenzoic acid)-di-tert-butylphenyl) propyl] amide (2), N-(4-hydroxyphenyl) amide 2-hidroxy benzoic acid (3), and 2-hydroxy-3-t-butyl-5-ethylbenzoic acid N-[3-(4-hydroxy-3,5-di-t-butylphenyl) propyl] amide (4). It is shown that amides exist both in the unbound state and in complexes with intra- and intermolecular hydrogen bonding. Free phenolic groups of amides undergo photolysis, which leads to the formation of a triplet state and phenoxyl radicals RO·, presumably due to the absorption of the second photon by the excited singlet state. Triplet-triplet annihilation and recombination ($kr \approx 2.3 \cdot 10^8$ L mol$^{-1}$ s) are the main channels for the decay of the triplet state and radicals RO·. UV irradiation of the compounds leads to the excitation of amide groups, and radical products are not formed due to ionization of the NH bond. The process of initiated UV oxidation of the model substrate (methylolcate) in the presence of amides 1–4 was compared with the known antioxidants (AO): dibunol (2,6-di-tert-butyl-4-methylphenol) (5) and $\alpha$-tocopherol (6-hydroxy-2,5,7,8-tetramethyl-2-phytyl chromane) (6). It has been shown that all amides of salicylic acid (I–IV) effectively inhibit the oxidation of methyl oleate, initiated by UV irradiation. The mechanism of the inhibitory effect of compounds has been established, which is associated with the possibility of direct interaction of phenols with free radicals (antiradical activity). Testing of antiradical activity of amides (I–IV), estimated by the method of chemiluminescence, made it possible to determine the range of reaction rate constants with peroxyl radicals RO$_2$· $k_7$ = (0.52–6.86) • $10^4$ m L$^{-1}$ s$^{-1}$. The interrelation of antioxidant properties of amides of salicylic acid (I–IV) and features of their structure is established. It was shown that the introduction of o-tert-butyl substituents and the separation of aromatic fragments by three methylene groups lead to a significant increase in antioxidant activity.

**Keywords:** salicylic acid amide, photolysis, phenyl radical, recombination, TR and UV spectroscopy, hydrogen bond, deuteration, UV-initiated oxidation, antioxidant, antiradical activity

---

# 1. Introduction

It is known that UV irradiation is a destructive factor that accelerates the oxidation of organic materials: polymers, rubbers, fuel, food, and technical oils. The effect of UV rays is the cause of erythema and skin burns, can cause DNA fragmentation, and leads to melanoma. The mechanism of the destructive effect of UV rays is associated with the activation of free radical oxidation processes as a result of the formation of singlet oxygen [1]. Salicylic acid and its esters, the characteristic feature of which is maximum absorption in the range of 300 nm [1–3], are used to protect the skin from ultraviolet irradiation (methyl salicylate) and as promising photostabilizers of polymeric materials [4] and dyes [5]. Some N-substituted amide derivatives of salicylic acid serve as inhibitors of oxidation in the photo-irradiation of substrates and exceed in the efficiency of known synthetic and natural antioxidants [4]. The photochemical reactions of salicylic and 4-sulfosalicylic acid have been studied [6–8]. Therefore, it is interesting to study the features of the photochemistry of new amide derivatives of salicylic acid and their antioxidant properties in comparison with known synthetic and natural antioxidants. For the compounds studied, similar studies have not been conducted previously (**Structure I**).

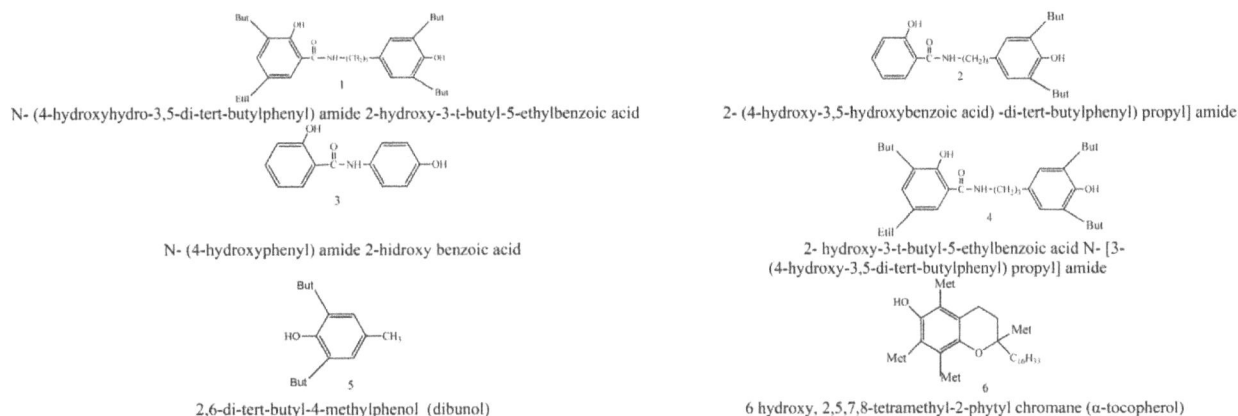

N- (4-hydroxyhydro-3,5-di-tert-butylphenyl) amide 2-hydroxy-3-t-butyl-5-ethylbenzoic acid

2- (4-hydroxy-3,5-hydroxybenzoic acid) -di-tert-butylphenyl) propyl] amide

N- (4-hydroxyphenyl) amide 2-hidroxy benzoic acid

2- hydroxy-3-t-butyl-5-ethylbenzoic acid N- [3-(4-hydroxy-3,5-di-tert-butylphenyl) propyl] amide

2,6-di-tert-butyl-4-methylphenol  (dibunol)

6 hydroxy, 2,5,7,8-tetramethyl-2-phytyl chromane (α-tocopherol)

**Structure I.** Structures of the test compounds.

# 2. Experimental

Optical spectra were recorded on an H P 8453 spectrophotometer (Hewlett Packard). IR spectra were measured on a Specord-75IR spectrophotometer. A laser stationary photolysis technique using excitation with a neodymium-doped yttrium aluminum garnet (Nd:YAG) laser (355 nm, pulse duration 5, bright area 0.03 cm$^2$, energy at a pulse 2 mJ (66 mJ cm$^{-2}$)

was used. The principal scheme of the system is similar to that described earlier [6]. The power of laser radiation was measured using a known procedure [9] with potassium ferrioxalate as a chemical actinometer.

Stationary photolysis of solutions of amides in heptane was carried out by a series of laser pulses for 4 minutes or by irradiation with a DRSh-250-3 mercury lamp (313–365 nm) in a closed cell (d = 0.4 cm) remote from the source and spherical mirror at equal distances (10 cm).

Deuteration was conducted by the dissolution of the compound in $CD_3OD$ followed by the evaporation of alcohol at 40–50°C.

Amides 1 and 2 were synthesized at the N. N. Vorozhtsov Novosibirsk Institute of Organic Chemistry (Siberian Branch of the Russian Academy of Sciences) and used without additional purification. The scheme of the synthesis and spectral characteristics of the products are published [9].

Carbon tetrachloride and heptane (reagent grade) were used for the preparation of solutions. All experiments were carried out at 298 K in a cell with an optical path length of 1 cm, except for specially indicated cases.

Oxygen was removed from solutions by argon bubbling. The kinetics of oxidation of methyloleate ($0.67$ mol $L^{-1}$) in the presence of amides of salicylic acid was studied in modified Warburg-type installations, fixing the amount of absorbed oxygen along the course of the reaction [4, 10]. As inert to the oxidation of the solvent, chlorobenzene purified by the simple distillation method was used. The temperature of the experiments is $(60 \pm 0.2)$°C [3, 10]. From the kinetic curves, the period of induction ($\tau$) was determined as a segment on the time scale, cut off by a perpendicular dropped from the point of intersection of the tangents. Initiation was carried out by preliminary irradiation of the substrate with a mercury lamp ($\lambda = 313$–$365$ nm) for 20 min at room temperature. The initiation rate was determined by the inhibitor method [11] using 6 as the control AO. The antiradical activity of salicylic acid amides was estimated from the rate constants in the elemental reaction with peroxyl radicals ($k_7$) by the chemiluminescent (CL) method with the initiation of cumene (isopropylbenzene) oxidation [12]. Chemiluminescence occurs as a result of the recombination of peroxyl radicals ($RO_2^\bullet$), since this reaction is very exothermic to excite luminescence in the visible region. When the antioxidant reacts with $RO_2^\bullet$, the intensity of chemiluminescence (CL) decreases to zero, which leads to a decrease in CL intensity. The consumption of antioxidant (AO) leads to an increase in the intensity of CL to the initial level. From the slope of the kinetic curves of CL, the rate of consumption of the antioxidant was determined [12]. The studies were carried out at a constant rate of generation of $RO_2^\bullet$ due to the thermal decomposition of azobisisobutyronitrile. The initiation rate was determined using a reference chromane inhibitor $C_1$ (2,2,5,7,8-pentamethyltocol). Oxidation of the substrate was carried out in a glass cell located in a light-tight chamber of a photometric block equipped with a photomultiplier FEU-29. The emitted light was focused on a photomultiplier using a system of spherical mirrors. To enhance the luminescence, 9,10-dibromanthracene was used in a concentration of $5 \cdot 10^{-4}$ M $L^{-1}$, which did not affect the kinetics of oxidation. The value of the constant $k_7$ was determined with the help of dependencies [12]:

$$\sqrt{I_0/I} = 1 + \left[ k_7 \times [InH]/\sqrt{k_6 \times W_i} \right] \qquad (1)$$

$$[d(I/I_0)/dt]_{max} = (0,22 \pm 0,02) \times k_7 \times \sqrt{W_i}/\sqrt{k_6} \qquad (2)$$

where I and $I_0$ are the luminescence intensity in the presence of the antioxidant and without it, respectively, Wi is the initiation rate, and $k_6$ is the rate constant of the chain termination in the known liquid-phase oxidation scheme of hydrocarbons [1].

To avoid the influence of antioxidant conversion products and the possibility of disturbance of the stationary process, the value was extrapolated $[d(I/I_0)/dt]_{max}$ to zero antioxidant concentration.

## 3. Results and discussion of IR spectra of native compounds for examples 1 and 2 and their photolysis products

It is known that salicylic acid and its derivatives in organic aprotic solvents form both intramolecular and intermolecular hydrogen bonds between the phenolic hydroxyl and the neighboring carbonyl group [13–16]. We note that the intramolecular hydrogen bond is predominantly formed at low concentrations and the intramolecular hydrogen bond at high concentrations [14–16]. During the presence of an intramolecular hydrogen bond in salicylic acid, the stability of monomeric complexes was confirmed by molecular modeling. A comparative analysis of IR spectra of salicylic and acetylsalicylic acids and methyl salicylate showed that the absorption band of OH groups participating in intramolecular hydrogen bond is 3230 cm$^{-1}$ and the complex spectrum in the range 2500–3300 cm$^{-1}$ characterizes the absorption of dimers [14–17]. The mechanism of formation of dimeric products from an intermediate reactive compound obtained as a result of disproportionation is discussed [17].

The formation of intra- and intermolecular hydrogen bonds was previously proven by several independent methods for the derivatives of salicylic acid [18–21].

The formation of an intramolecular hydrogen bond for salicylic aldehyde and o-nitrophenol (-O-H ... O = C- and -O-H ... O = N-, respectively) was demonstrated. This coupling exhibits an intense maximum at 3200 cm$^{-1}$ and a fairly strong shift of the v (OH) band to low frequencies. The vibration band v (OH) on an intermolecular hydrogen bond usually has a complex structure with a maximum in the 3400–3560 cm$^{-1}$ range and is characterized by a half-width of ~ 400 cm$^{-1}$ (see 10 and 13). Intermolecular hydrogen bond (O-H ... O =) between phenol (a-tocopherol) and quinone (ubiquinone Q10) appears in the IR spectrum as a band with a maximum at 3545 cm$^{-1}$ (see 14) [22].

To reveal specific features of the molecular structures of amides 1 and 2, we examined the IR spectra of solutions in CC1$_4$ in a range of 1600–4000 cm$^{-1}$ (**Figure 1**). The indicated range contains the band of stretching vibrations of the unbound phenol group (v(OH) = 3644 cm$^{-1}$) and the band with a maximum at 3454 cm$^{-1}$, which is due to the presence of the v(NH) amide group usually observed near 3450 cm$^{-1}$. For stationary photolysis of amide 1 in CC1$_4$, the IR spectra exhibit a decrease in the intensity of stretching vibration bands of isolated OH

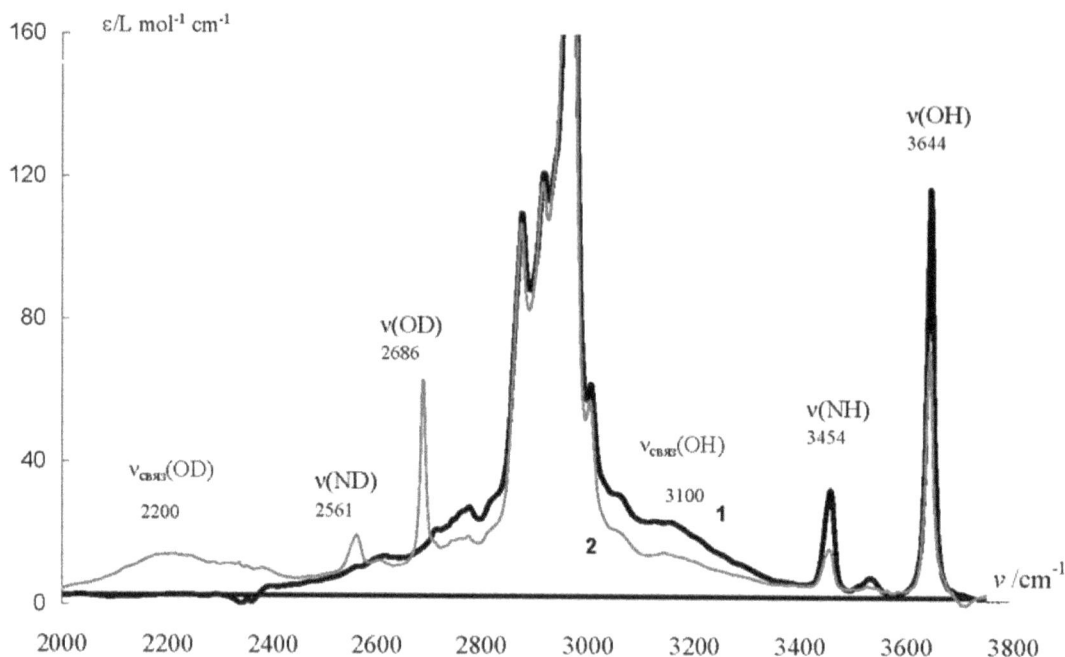

**Figure 1.** IR spectra of amide **1** ($10^{-2}$ Mol $L^{-1}$) in CCl$_4$: (1) initial compound and (2) initial compound partially deuterated at the OH and NH groups.

($v$(OH) = 3644) cm$^{-1}$ and NH groups ($v$(NH) = 3454) cm$^{-1}$ (**Figure 2**). Upon the irradiation of the solution for more than 3 min, a band with a maximum at ~3424 cm$^{-1}$ assigned possibly to the absorption band of the product appears and increases in the spectrum of amide **1**.

We studied the possibility of formation of hydrogen bonds of different natures in structures **2** and **3**. The IR spectra of amides **1** and **2** exhibit a broad complicated band at 2300–3400 cm$^{-1}$. According to literature data, the phenol group ($v$(OH)) involved in intra- and intermolecular hydrogen bonds can absorb in this range. However, the identification of hydrogen bonds is impeded by the fact that an intense absorption of =C-H and C-H bonds is observed in a range of 2800–3100 cm$^{-1}$. The most intense absorption concerns just this range in the spectra studied. It is noteworthy that a broad doublet band with maxima at 3045 and 3150 cm$^{-1}$ appears in the spectrum at a shoulder of the absorption band of the =C-H and C-H bonds in a range of 2300–3400 cm$^{-1}$.

To reveal the question about the existence and nature of hydrogen bonds, amides **1** and **2** were deuterated, and then a comparative analysis of the spectra of the initial and partially deuterated molecules was carried out. It is seen upon the superposition of the indicated spectra (for amide **1**; see **Figure 1**) that the absorption bands of the unbound OH (3644 cm$^{-1}$) and NH (3454 cm$^{-1}$) groups do not change their position, but their intensity decreases. At the same time, new bands assigned to the unbound OD and ND groups (frequencies 2686 and 2561 cm$^{-1}$, respectively) appear in the spectrum of the deuterated molecules. The low-frequency spectral range contains a new doublet band with maxima at 2250 and 2175 cm$^{-1}$. The appearance of this band indicates hydrogen OD bonds and is due to the resonance Fermi interaction and totally difference transitions involving low-frequency vibrations of the D bond [23]. It is known that the ratio between positions of the bands of the OH groups involved in the

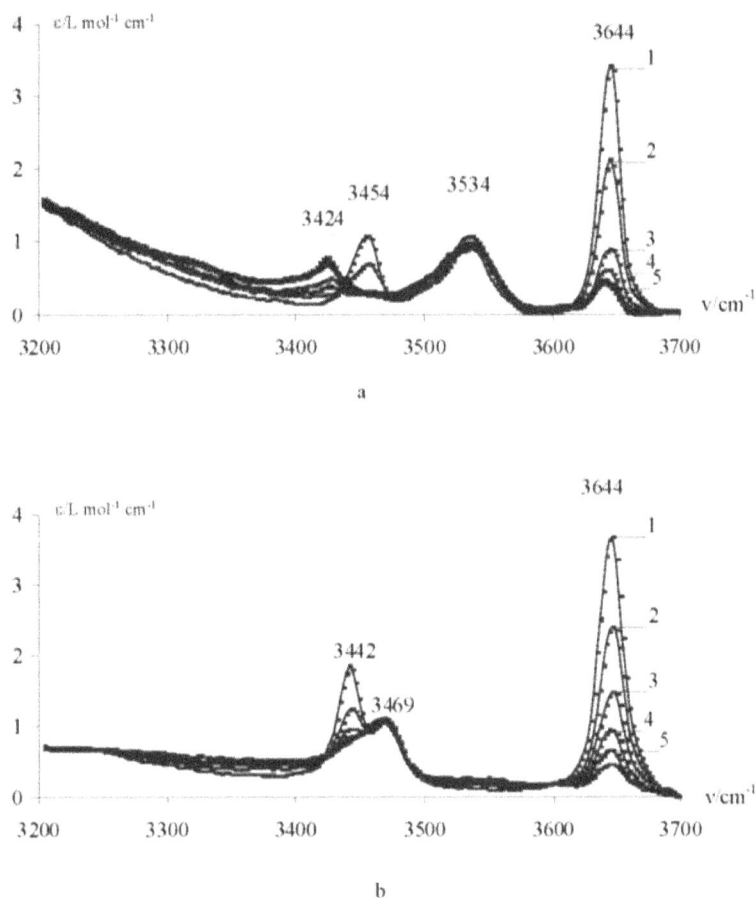

**Figure 2.** IR spectra for stationary photolysis of salicylic acid amides **1** (a) and **2** (b) ($5 \cdot 10^{-3}$ Mol $L^{-1}$) in $CCl_4$ after 1, 2, 3, 4, and 5 min (curves 1–5, respectively) of irradiation.

formation of structures with a hydrogen bonds and the OD group in a similar complex is $v_c(OH) \approx \sqrt{2}v_c(OD)$. It follows from the calculation by this formula that in the initial non-deuterated structure the OH groups involved in the hydrogen bond absorb at 3045 and 3150 $cm^{-1}$. These results coincide with experimental data: as indicated above, two maxima are observed in this range at the shoulder of the intense absorption of the =C-H and C-H groups. As a result of deuteration, the intensity of this doublet decreases, which additionally proves the presence of hydrogen bonds. The band intensity remains nearly unchanged with the temperature change in a range of 25–70°C, indicating a high strength of the formed intra- and intermolecular hydrogen bonds.

The change in the concentration of amides 2 and 3 in the range $1.0 \cdot 10^{-2}$–$25 \cdot 10^{-4}$ $mol^{-1}$ did not affect the shape of IR spectra. To correctly compare the intensities of the absorption bands of solutions of different concentrations, they were normalized per unit concentration and unit thickness of the absorbing layer:

$$\varepsilon = \frac{\ln{I_0/I}}{c \cdot l} \qquad (3)$$

where C is the concentration of the substance in the solution and l is the thickness of the absorbing layer. As a result, there was an insignificant decrease in the intensity of the absorption bands v (= C-H, -C-H) and v (OH), referred to complexes with hydrogen bond, which also indicates the strength of the formed complexes.

It is known that amides and secondary amines are weaker proton donors in hydrogen bonding than hydroxyl-containing compounds. The v(NH) band is shifted over the band of the monomer from 14 to 74 cm$^{-1}$ upon the formation of a complex of the N-H...O=C- type [10]. It is seen from **Figure 1** that no substantial shift of the v(NH) bands is observed in the spectral of the studied compounds. It is most likely that no hydrogen bond is formed at the amide groups and carbonyl of the N-H...O=C type in amides **1** and **2**.

Thus, an analysis of the spectral absorption distribution suggests that the following structures are most probable for the studied amides. Amides **1** and **2** exist predominantly as complexes with the structure I) and intermolecular hydrogen bond (**Structure II**) formed by the phenol and carbonyl groups. The phenolic OH groups arranged in the adjacent position with the C = 0 group are involved in complexes. It is most probable that sterically hindered phenol groups are free in both complexes with an intramolecular hydrogen bond and in structures with an intermolecular hydrogen bond. Therefore, our investigations were continued by the study of the kinetics of phototransformation of the studied amides at the absorption bands of the unbound v(OH) and v(NH) groups.

Analysis of IR spectra for stationary photolysis of amides **1** and **2**. An analysis of stretching vibrations for stationary photolysis of amides **1** and **2** shows that the intensity of the band v (OH) = 3644 cm$^{-1}$ regularly decreases under UV radiation. These changes indicate the ionization of free phenolic hydroxyls.

The intensity of the band in the range of stretching vibrations of the NH group (v(NH) 3454 cm$^{-1}$ (1) and 3442 cm$^{-1}$ (2) decreases with an increase in the UV irradiation time (see **Figure 2**). The band with a maximum at 3424 cm$^{-1}$ which can be attributed to the v(NH) vibration of long-lived intermediates appears and increases in the spectrum of amide **1** (see **Figure 2a**). For amide **2**, the indicated band is smoothened and exists as a shoulder (see **Figure 2b**).

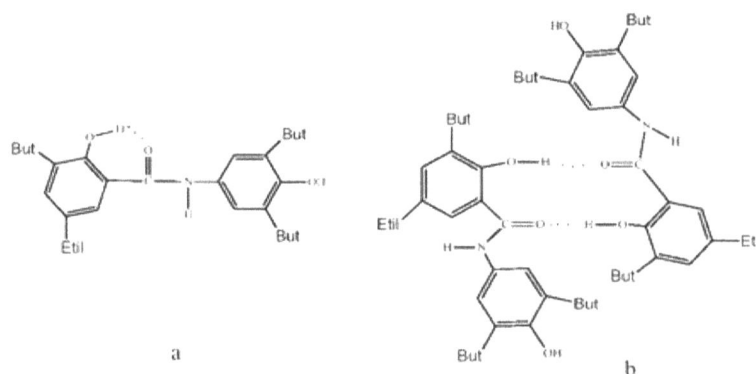

**Structure II.** Intramolecular (a) and intermolecular (b) hydrogen bonds, on the model, amide structure 4.

The position and intensity of bands at 3534 and 3469 cm$^{-1}$ remain almost unchanged after UV irradiation. They can be assigned to stretching vibrations of the C-H bonds of the aromatic fragments of amides **1** and **2**, respectively. The bathochromic shift of the indicated component of the spectrum related to a higher degree of delocalization is observed for amide **1**. In structure 1 both benzene rings, amide group, and phenolic hydroxyls are involved in the general system of conjugation, whereas in structure 2, the residue of salicylic acid amide and the fragment of sterically hindered phenol are separated by three methylene groups.

An analysis of the IR spectra allows one to conclude that during photolysis amides transform into the excited state, but no radical products are formed upon the cleavage of the N-H bond. Free phenolic hydroxyls that do not participate in hydrogen bonding undergo ionization. Thus, the formation of phenoxyl radicals should be expected for the photolysis of the studied structures. Most likely, no radical products at the amide groups are formed.

## 4. Analysis of UV: Vis spectra for stationary photolysis of amides 1 and 2

The UV spectrum of amide **1** (**Figure 3a**) exhibits absorption maxima at 225 and 325 nm responsible for the $\pi$-$\pi^*$ and n-$\pi^*$ transitions, respectively. For stationary photolysis (irradiation with a mercury lamp, 313–365 nm), the intensity of the absorption bands at 325 and 225 nm decreases, and new bands with maxima at ~255 and ~295 nm appear in the spectrum. An absorbance at 370–400 nm in the form of a shoulder also increases.

These new absorption bands belong to the products of phototransformations of amide **1**. The band of the $\pi$-$\pi^*$ transition (225 nm) undergoes the bathochromic ("red") shift ($\Delta\lambda \approx 25$ nm), and the band of the n-$\pi^*$ transition (325 nm) experiences by the hypsochromic ("blue") shift ($\Delta\lambda \approx 39$ nm) (see **Figure 3a, b**).

In amide **2** (see **Figure 3b**), the bands of electron transitions of the native and phototransformed molecules experience the hypsochromic shift compared to amide 1 caused by a lower degree of electron density delocalization. After irradiation, the intensity of the absorption bands of molecule 2 at 311 nm (n-$\pi^*$ transition) and 220 nm ($\pi$-$\pi^*$ transformation) decreases, and new bands with maxima at ~250 and ~ 300 nm and a shoulder in a range of 330–400 nm appear in the spectrum.

The pattern of spectral change in solutions of amide 1 in heptane upon irradiation with pulses of a neodymium laser (for 4 min) is nearly identical to that observed upon lamp photolysis (**Figure 4a**). The UV spectra also demonstrate a decrease in the absorption band intensity of the initial compound at 325 nm (band of the n-$\pi^*$ transition) and 226 nm (band of the $\pi$-$\pi^*$ transition) and the appearance of new bands with maxima at 255, 285, and 370 nm, which can be assigned to photoproducts (**Figure 4a, curve 5**). The band of the $\pi$-$\pi^*$ transition experiences the bathochromic ("red") shift ($\Delta\lambda \approx 9$ nm), whereas the band of the n-$\pi^*$ transition undergoes the hypsochromic ("blue") shift ($\Delta\lambda \approx 39$ nm).

These new absorption bands belong to the products of phototransformations of amide **1**. The band of the $\pi$-$\pi^*$ transition (225 nm) undergoes the bathochromic ("red") shift, and the band

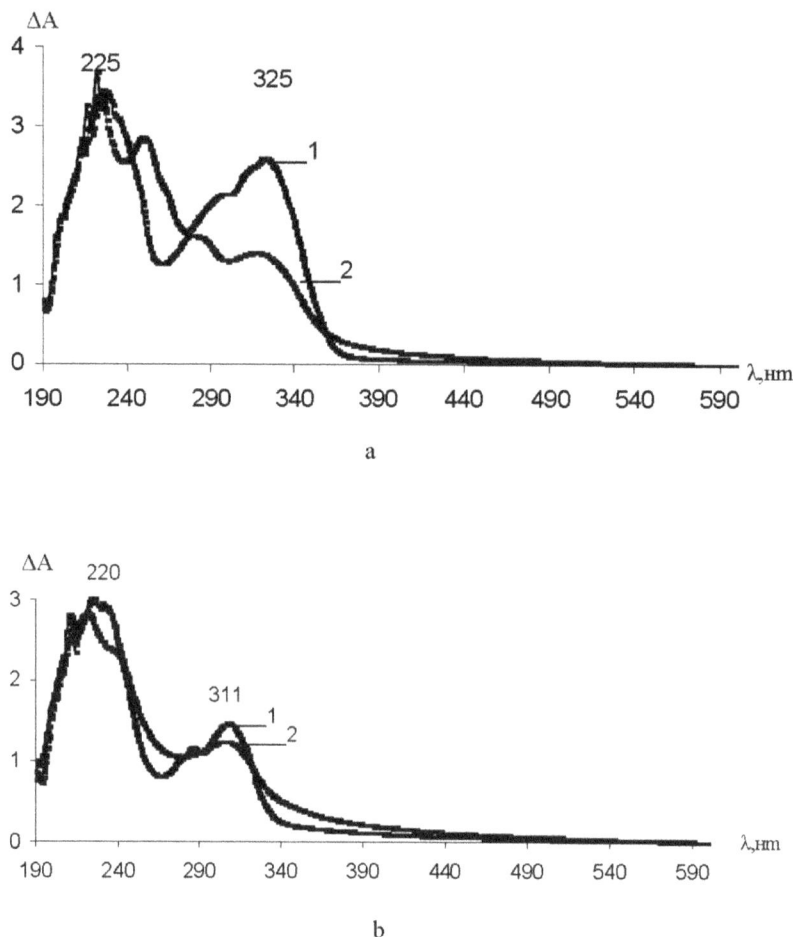

**Figure 3.** Optical absorption spectra for stationary photolysis of salicylic acid amides 1 (a) and 2 (b) (5 • 10⁻⁴ mol L⁻¹) before (1) and after (2) irradiation in heptane for 5 min.

($\Delta\lambda \approx 25$ nm) of the n-$\pi^*$ transition (325 nm) experiences by the hypsochromic ("blue") shift (see **Figure 3a, b**) ($\Delta\lambda \approx 39$ nm).

In amide **2** (see **Figure 3b**), the bands of electron transitions of the native and phototransformed molecules experience the hypsochromic shift compared to amide **2** caused by a lower degree of electron density derealization. After irradiation, the intensity of the absorption bands of molecule 2 at 311 nm (n-$\pi^*$ transition) and 220 nm ($\pi$-$\pi^*$ transformation) decreases, and new bands with maxima at ~250 and ~ 300 nm and a shoulder in a range of 330–400 nm appear in the spectrum.

The pattern of spectral change in solutions of amide **2** in heptane upon irradiation with pulses of a neodymium laser (for 4 min) is nearly identical to that observed upon lamp photolysis (**Figure 4a**). The UV spectra also demonstrate a decrease in the absorption band intensity of the initial compound at 325 nm (band of the n-$\pi^*$ transition) and 226 nm (band of the $\pi$-$\pi^*$ transition) and the appearance of new bands with maxima at 255, 285, and 370 nm, which can be assigned to photoproducts (**Figure 4a, curve 5**). The band of the $\pi$-$\pi^*$ transition experiences the bathochromic ("red") shift ($\Delta\lambda \approx 9$ nm), whereas the band of the n-$\pi^*$ transition undergoes the hypsochromic ("blue") shift ($\Delta\lambda \approx 39$ nm).

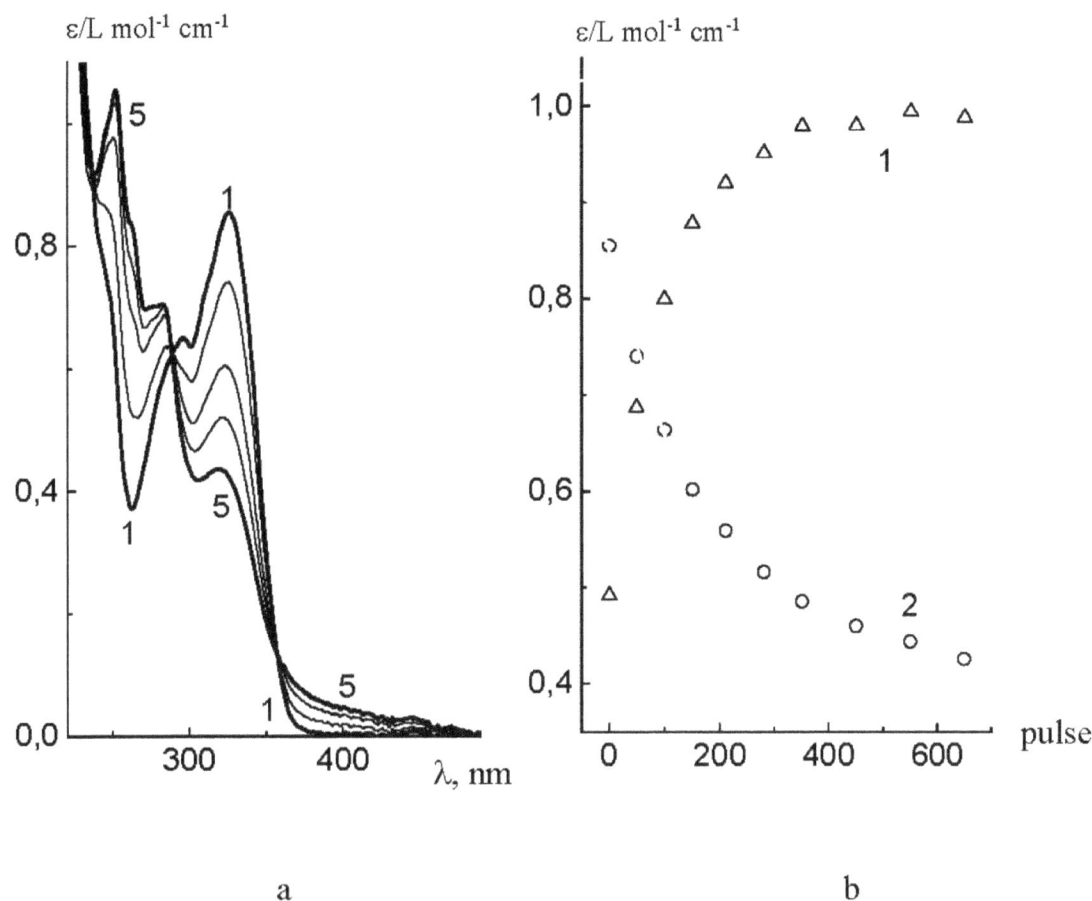

**Figure 4.** (a) Optical absorption spectra of a solution of amide 1 ($3.45 \cdot 10^{-4}$ Mol L$^{-1}$) in heptane after 0 (1), 50 (2), 110 (3), 180 (4), and 250 (5) laser pulses. (b) Change in the absorbance at the wavelengths 255 (1) and 325 nm (2) during photolysis.

It should be mentioned that during photolysis of amide **1** isosbestic points at 238, 290, and 357 nm are retained, indicating a low photoactivity of the final products compared to the initial compound. The mentioned qualitative changes in the spectra of the substances under irradiation with a mercury lamp and a neodymium laser indicate the phototransformation of molecules accompanied by the formation of fairly stable products with the electron density redistribution in the system of $(\pi\text{-}\pi^*)$ and $(n\text{-}\pi^*)$ conjugation.

We also studied the dependence of the absorbance of amide 1 at 255 and 325 nm attributed to the photolyzed and native molecules, respectively, on the number of laser pulses fed to the sample (see **Figure 4b**). It is seen from **Figure 4b** that after 650 pulses 100% of the initial compound transform into photolysis products.

The combined interpretation of the data on pulse and stationary photolysis suggests the structure of the main final photochemical products and reactions of its formation (**Scheme 1**).

The formation of the compound with the quinoid structure (iminoquinone) corresponds to the disappearance of stretching vibration bands of the OH and NH groups (see **Figure 2**) and the disappearance of the absorption bands at 240–290 nm characteristic of quinones and iminoquinones [23–24]. The presence of bulky substituents in the *ortho-* and *para*-positions of

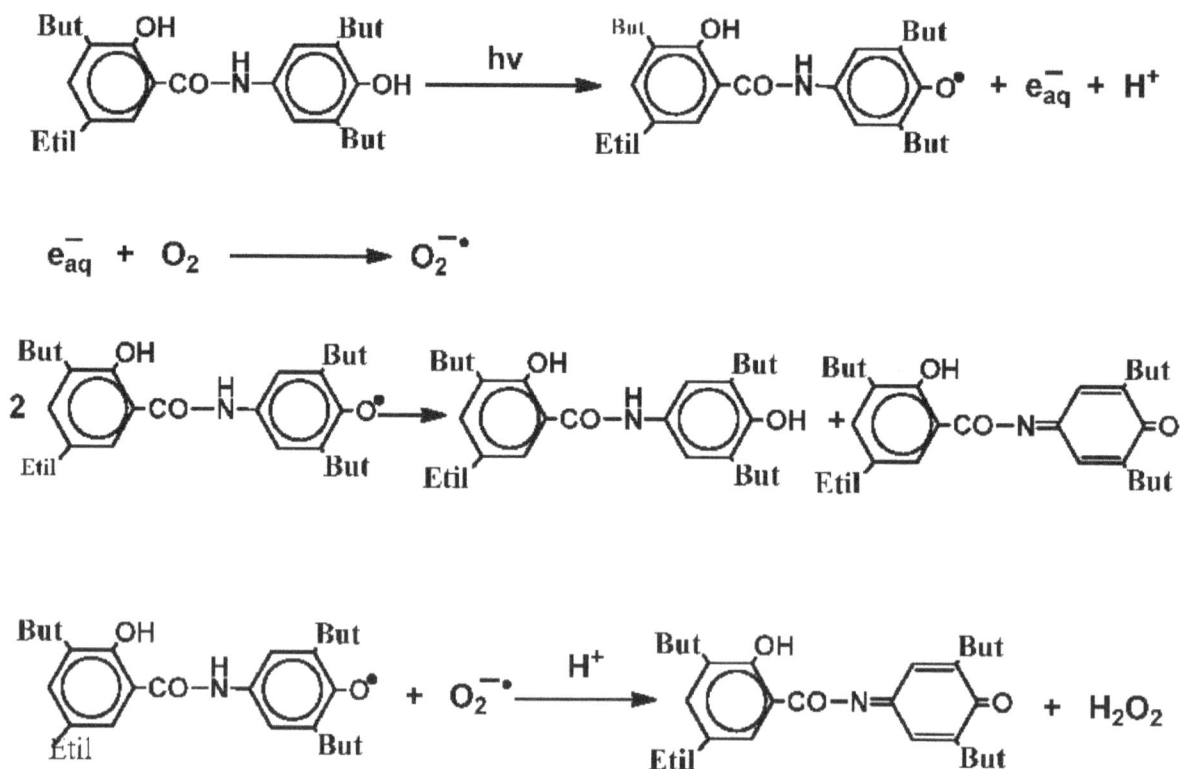

**Scheme 1.** The formation reactions and structure of the main products of the photochemical conversion of salicylic acid amides.

the iminophenol fragment prevents the recombination of phenoxyl radicals with the formation of dimeric products.

Laser pulse photolysis of amide **1**. The excitation by a laser pulse (308 nm) of deoxygenated aqueous solutions of salicylic acid amides results in an intermediate absorbance consisting of two bands with maxima at 380 and 510 nm (**Figure 5a**), which disappear with significantly different rates (**Figure 5b**). These data indicate the formation of several intermediate species after a laser pulse. The lifetime of the band at 510 nm decreases substantially in the presence of oxygen, which indicates that the band belongs to the absorbance of amide 1 from the triplet T: state. The main channel of triplet state decay is triplet-triplet annihilation. It is known that the triplet-triplet absorption band of salicylic acid in cyclohexane has a maximum at 440 nm.[1] The bathochromic shift of the absorption band of amide **1** at 70 nm is caused, most likely, by an additional conjugation due to the introduction of the phenol substituent at the amide group of salicylic acid and to other types of the solvent.

In the presence of oxygen, the triplet state of amide **1** rapidly disappears during quenching, which makes it possible to detect one more, longer-living intermediate, whose optical spectrum consists of two absorption bands with maxima at 480 and 380 nm (**Figure 6a**). The kinetics of the disappearance of this absorbance is presented in **Figure 6b**. The absorbance amplitude at 380 nm ($\Delta A^{380}$) depends nonlinearly on the laser pulse intensity (**Figure 7a**) and can be expressed by the following equation:

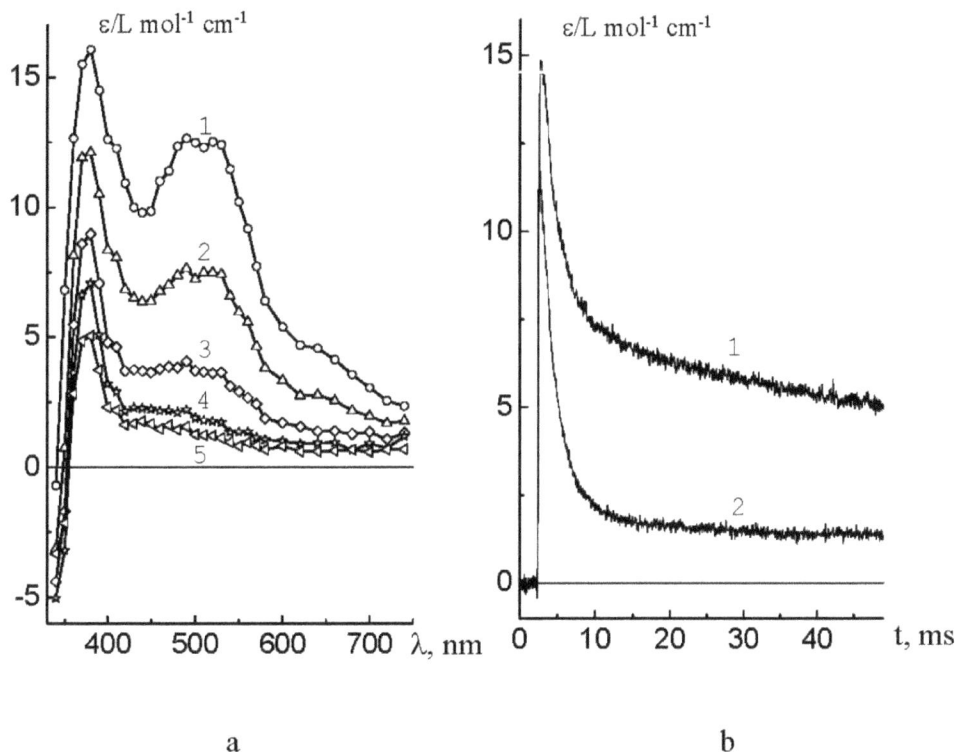

**Figure 5.** (a) Intermediate absorption spectra for stationary photolysis of a deoxygenated solution of amide 1 (3.45 • $10^{-4}$ Mol $L^{-1}$) in heptane; 1–5 spectra in 0, 1.6, 4, 10, and 48 |is after a laser pulse (2.2 mJ pulse$^{-1}$), respectively. (b) Kinetic curves of a change in the absorbance at 380 (1) and 510 nm (2).

$$\Delta A^{380} = 1.6 \times 10^{-4} \times I + 1.4 \times 10^{-5} \times I^2 \qquad (4)$$

where $I$ is the laser pulse intensity, mJ cm$^2$. These data indicate that the long-lived intermediate is formed in both one- and two-quantum processes. The product of the quantum yield of the one-quantum process (cp) by the absorption coefficient of the long-lived intermediate at 380 nm ($g^{380}$) can be estimated as ($g^{380}$ = 80 mol $L^{-1}$) cm$^{-1}$:

$$\text{ROH} - \text{hv} \rightarrow \text{RO}^\bullet + e^-_{solv} + \text{H}^+ \qquad (5)$$

Phenoxyl radical can also be formed upon the absorption of the second quantum of light by the excited singlet or triplet states of phenols. For example, two-quantum photoionization for the absorption of the second quantum by the excited singlet state of these compounds to form a pair hydrated RO$^\bullet$ electron-organic radical was observed [5–7] in the study of photochemistry of aqueous solutions of salicylic and sulfosalicylic acids. These data suggest that the long-lived intermediate observed for photolysis of amide 1 is the phenoxyl radical formed in the one- (3) and two-quantum processes **(4)**:

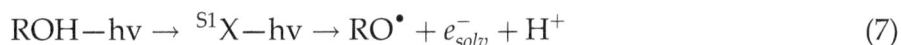

$$^{S1}\text{X} - \text{hv} \rightarrow \text{RO}^\bullet + e^-_{solv} + \text{H}^+ \qquad (6)$$

$$\text{ROH} - \text{hv} \rightarrow \, ^{S1}\text{X} - \text{hv} \rightarrow \text{RO}^\bullet + e^-_{solv} + \text{H}^+ \qquad (7)$$

The solvated electron in heptane absorbs in the IR range (a maximum at 1600 nm [8]) and cannot be detected using the system used in the work. It should be mentioned that the

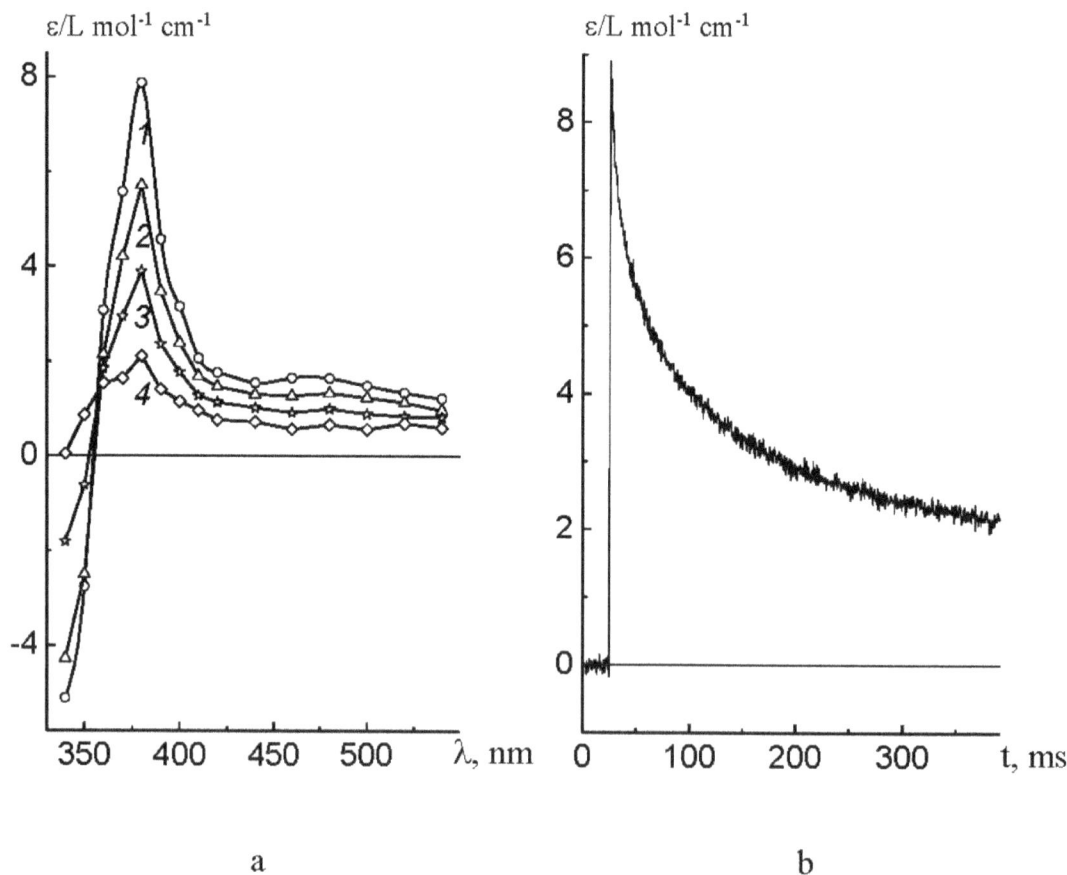

**Figure 6.** (a) Intermediate absorption spectra for stationary photolysis of a solution of amide 1 ($3.45 \cdot 10^{-4}$ Mol $L^{-1}$) in heptane at an oxygen concentration in the solution of $3 \cdot 10^{-3}$ Mol $L^{-1}$ (**1–4**) spool ra in 3, 12, 90, and 380 (is after a laser pulse, respectively. (**b**) Kinetic curve of a change in the absorbance at 380 nm. It is known that one of the intermediate products of photoionization of phenols (ROH) is the corresponding phenoxyl radical (see refs. 17 and 18) Formed in the reaction.

absorption band maxima of radical RO$^\bullet$ of amide 1 (380 and 480 nm) are shifted to the red range compared to the unsubstituted phenoxyl radical (290 and 400 nm) [25, 26]. This is due to the iminophenol substituent in molecule 1. It is known that the introduction of aromatic substituents results in the bathochromic shift of the absorption bands of phenoxyl radicals, in particular, ongoing from the unsubstituted phenoxyl radical to radicals of 4-phenylphenol and 4,4′-biphenol, the long wavelength band maximum shifts from 400 nm by 560 and 620 nm, respectively [27].

It should be mentioned that phenoxyl radicals decay predominantly in recombination reactions [26, 27]. The kinetics of the disappearance of the absorbance of radical RO$^\bullet$ (380 nm) is described rather well by the second-order law. The linear dependence of the observed rate constant ($k_{app}^{380}$) on the absorbance amplitude (**Figure 7b**) makes it possible to determine the ratio of the recombination rate constant of phenoxyl radicals, $2k_{pek}/\varepsilon^{380} = 1.6 \cdot 10^5$ cm s$^{-1}$. The kinetics of radical decay was determined in solutions containing oxygen (to accelerate the disappearance of the absorbance of the triplet state of amide 1), and, therefore, the section cut in the ordinate (see **Figure 7b**) corresponds, most likely, to the reaction of RO$^\bullet$ with oxygen. Under normal conditions, the concentration of oxygen in a heptane solution is $3 \cdot 0^{-3}$ mol $L^{-1}$ [24], which makes it possible to estimate the rate constant of the reaction with oxygen,

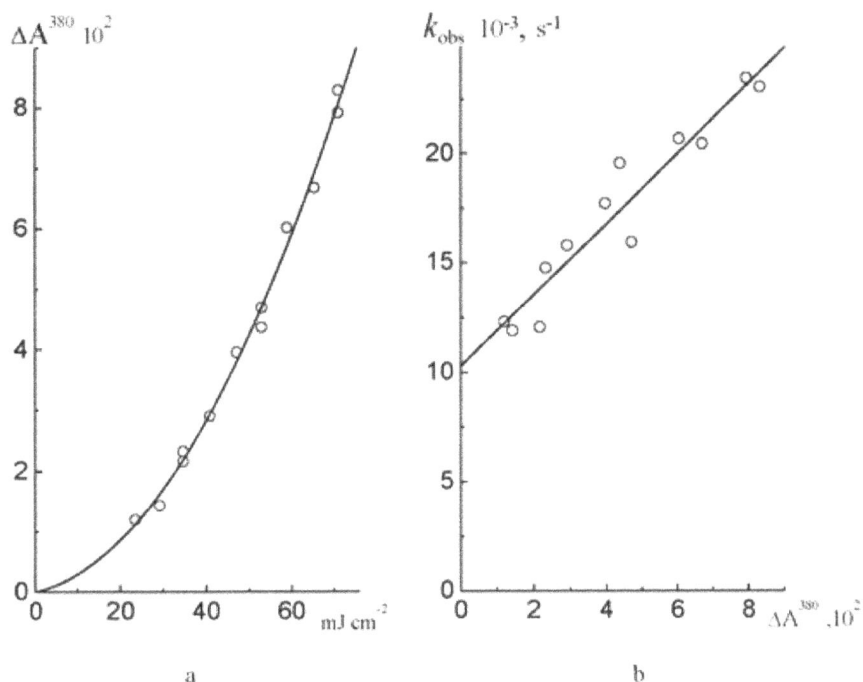

**Figure 7.** (a) Absorbance amplitude ($\Delta a^{380}$) и laser pulse intensity (the initial absorbance at the wavelength 355 nm is equal to 0.44). (b) Reaction rate constant ($A'_{obs}$) for the disappearance of the absorbance of radical $RO^\bullet$ of amide 1 ($3.45 \bullet 10^{-4}$ Mol $L^{-1}$) from signal amplitude at 380 mil.

$k_{Q2} \approx 3.4 \cdot 10^6$ L mol$^{-1}$ s$^{-1}$. The value of rate constant is more than three orders of magnitude lower than the diffusion rate constant in this solvent. Low rate constants for the reaction with oxygen are characteristic of phenoxyl radicals [19, 27]:

$$A = A_0 + \frac{\phi_{app} N_{abs}(A_\infty - A_0)}{N_a C_X V} N_{pulse} \qquad (8)$$

At the initial stage of photolysis, the change in the absorbance of the sample can be expressed by the equation where $\varphi_{app}$ is the apparent quantum yield of photolysis of amide 1 at a given intensity of a laser pulse, $N_{abs}$ is the number of quanta absorbed by the sample, $N_a$ is Avogadro's number, $V$ is the volume of the sample, $N_{pulse}$ of laser pulses, $C_1$ is the initial concentration of amide 1, and $A_0$ and $A_\infty$ are the initial and final absorbances of the sample. The latter value can be estimated from the data of **Figure 4b** assuming 100% conversion of amide 1 to photolysis products. Accordingly, knowing the value of $A_\infty$, one can estimate the apparent quantum yield of photolysis of amide **1** by Eq. (8). For a laser pulse intensity of 60 mJ cm-$^2$, the apparent quantum yield of photolysis of amide 1 is 0.09.

Assuming that at this intensity the main mechanism of $RO^\bullet$ decay is recombination, the absorption coefficient of $RO^\bullet$ at a wavelength of 380 nm and, correspondingly, the recombination constant of radicals can be estimated from the value of the signal of intermediate absorbance at 380 nm at the given intensity. The obtained value of the absorption coefficient of phenoxyl radicals of 1 ($\varepsilon^{380} \approx 2.9 \bullet 10^3$ cm s$^{-1}$) is close to that presented for $RO^\bullet$ of salicylic acid ($\varepsilon^{390} = (2.5 + 0.3) \bullet 10^3$ cm s$^{-1}$ in an aqueous solution).[5] It is known that the absorption coefficients of phenoxyl radicals depend slightly on their structure and in a range of 2.9–4.0 • $10^3$ cm s$^{-1}$. The recombination rate constant of RO$^\bullet$ radicals of 1 was $2k_r \approx 4.6 \bullet 10^8$ L mol$^{-1}$ s$^{-1}$. The obtained value corresponds to literature recombination constants for phenoxyl radicals that form relatively unstable dimers [28]. The recombination rate of unsubstituted salicylic acid is substantially higher than $2k_t \approx (1.8 + 0.3) \bullet 10^9$ L mol$^{-1}$ s$^{-1}$ (in an aqueous solution) [7], and the order of magnitude of the rate constant is typical of the bimolecular decay of phenoxyl radicals [24, 29]. The decrease in the activity of phenoxyl radicals of UV-substituted amides in dimerization processes is due to steric factors and the use of other solvents.

Thus, UV irradiation of UV-substituted salicylic acid amides organized in complexes with intra- and intermolecular hydrogen bonds induces the phototransformation of free phenolic hydroxyls with the formation of phenoxyl radicals RO$^\bullet$, which decay in recombination following the second-order law with the rate constant $k_T \approx 2.3 \cdot 10^8$ L mol$^{-1}$ s$^{-1}$. The NH groups undergo excitation, but no radical products are formed. The dimerization products of amides 1 and 2 are relatively stable under the photolysis conditions. With phenoxy radicals, oxygen reacts at a relatively low rate [29, 30].

# 5. Study of the kinetics of inhibition of the oxidation process by N-substituted amides of salicylic acid

In the present work, the inhibitory and photostabilizing properties of compounds that are amide derivatives of salicylic acid, characterized by the conjugation of the electron density and the degree of spatial screening of the phenolic hydroxyls (structure 1) are investigated.

It is known that salicylic acid and its derivatives are capable of absorbing UV rays in the range of 301–305 nm [3, 8]. It can be assumed that the modified structures studied by us can also absorb UV rays in this range, which is dangerous for the development of skin cancer. The compounds can potentially exhibit antioxidant and antiradical activity, due to the presence of two phenolic groups in their chemical structure.

It is known that ultraviolet irradiation leads to the formation of singlet oxygen $O_2^*$, which joins the double bond of unsaturated lipids [11, 12]. As a result of further isomerization of cyclic peroxides and photochemical decomposition of hydroperoxides, hydroxyl (OH$\bullet$) and alkoxy radicals (rO$\bullet\cdot$) are formed which react with the substrate RH in an oxygen atmosphere to form peroxyl radicals RO$_2\bullet$ (**Scheme 2**).

To study the possibility of photostabilization of oxidation processes caused by the action of UV rays, it was considered necessary to study the electronic spectra of N-substituted amide derivatives of salicylic acid (**1–4**) in comparison with **5** (**Figure 3**). The presence of intense absorption bands in the UV range (190–350 nm) for all amides of salicylic acid has been established. The most effective UV rays absorb compounds 1 and 2. This absorption is absent in **5**. Thus, the absorption of UV rays that initiate the process of lipid oxidation is one of the mechanisms of photostabilizing action of N-substituted amides of salicylic acid (**Figure 8**).

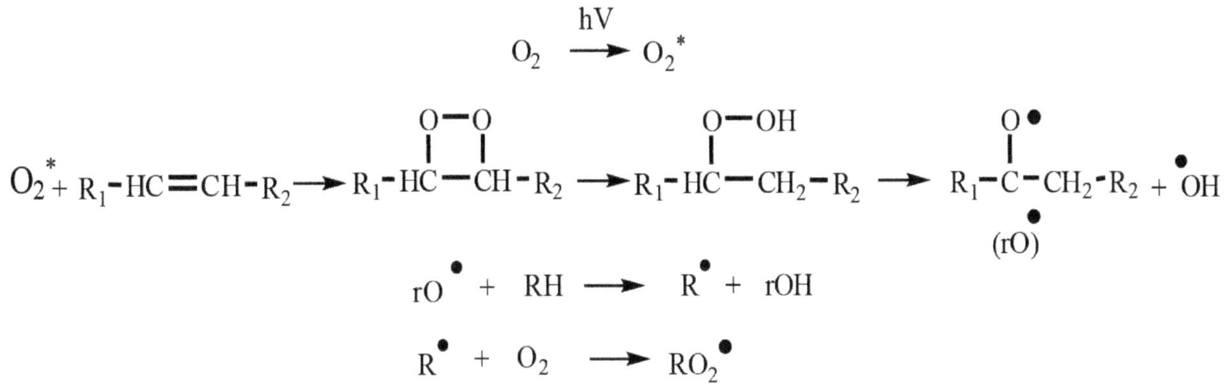

$$O_2 \xrightarrow{hV} O_2^*$$

$$O_2^* + R_1\text{-}HC=CH\text{-}R_2 \longrightarrow R_1\text{-}\overset{\overset{\displaystyle O\text{---}O}{|}}{HC}\text{-}\overset{\overset{\displaystyle |}{}}{CH}\text{-}R_2 \longrightarrow R_1\text{-}\overset{\overset{\displaystyle O\text{---}OH}{|}}{HC}\text{-}CH_2\text{-}R_2 \longrightarrow R_1\text{-}\overset{\overset{\displaystyle O^\bullet}{|}}{C}\text{-}CH_2\text{-}R_2 + {}^\bullet OH$$

$$(rO)$$

$$rO^\bullet + RH \longrightarrow R^\bullet + rOH$$

$$R^\bullet + O_2 \longrightarrow RO_2^\bullet$$

**Scheme 2.** Scheme of peroxyl radical $RO_2^\bullet$ formation during UV irradiation during the reaction of singlet oxygen $O_2^*$ with unsaturated compounds.

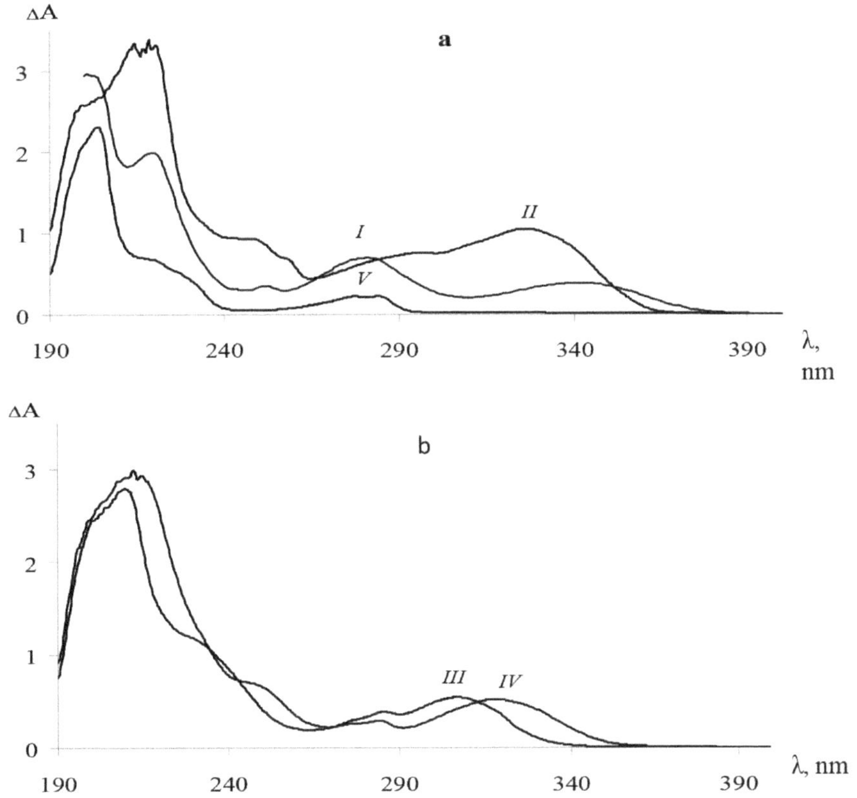

**Figure 8.** UV spectra of amides: (a) I is 3, II is 4, and V is 5; (b) III is 2, IV is 1, and in heptane $c = 5 \cdot 10^{-4}$ Mol $L^{-1}$, $d = 0.2$ cm.

The mechanism of the complex multistage process of free radical oxidation of the substrate is described by a conventional scheme [11, 12]. The oxidation process is carried out by.

peroxyl radicals reacting with a substrate with a rate constant $k_2$ [29].

$$RO_2^\bullet + RH \xrightarrow{k_2} ROOH + R^\bullet \qquad (9)$$

In the presence of an antioxidant (InH), free radicals are killed according to reaction 7 to the conventional scheme [24, 25].

$$RO_2^{\bullet} + InH \xrightarrow{k_7} ROOH + In^{\bullet} \tag{10}$$

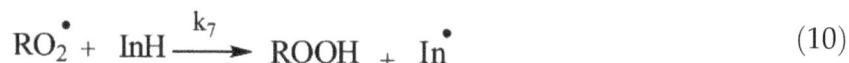

In the presence of an antioxidant, induction periods ($\tau$) appear, and the magnitude of which depends on their number. In this paper, the photostabilizing effect of N-substituted salicylic acid derivatives, differing in electron density conjugation and the degree of spatial screening of phenolic hydroxyls (structure 1), was studied. The antioxidant activity of amides of salicylic acid was evaluated by the value of induction periods ($\tau$) with UV-induced oxidation of the model substrate, methyloleate. The kinetic curves of oxygen absorption by methyl oleate in the presence of test compounds are shown in **Figure 9**. As can be seen from **Figure 9**, all the amides (1–4) studied effectively inhibit the oxidation of the substrate. With comparable concentrations of AO, the magnitude of induction periods provided by different amides is significantly different. Differences in the inhibitory effect of compounds are determined by the peculiarities of their chemical structure. The most effective oxidation inhibitors are amides 1 and 2, which are sterically hindered diatom phenols, in the structure of which the conjugated fragments are separated by three methylene groups. In this case, both phenolic groups can interact with free radicals independently of each other. The presence of bulk tert-butyl substituents prevents the possibility of adverse reactions that reduce the effectiveness of the antioxidant. Compared with amides **1** and **2**, amides **3** and **4** exhibit a lower inhibitory effect.

In these compounds, both aromatic moieties form a general conjugation system with an amide group. The difference between amides **3** and **4** is the degree of spatial screening of phenolic groups (-OH). In structure **3** there is no tert-butyl substituents, which significantly increases the activity of oxidation products of phenols in the side processes of oxidation, in particular, in reaction with the substrate. This reaction leads to the continuation of the oxidation chains and

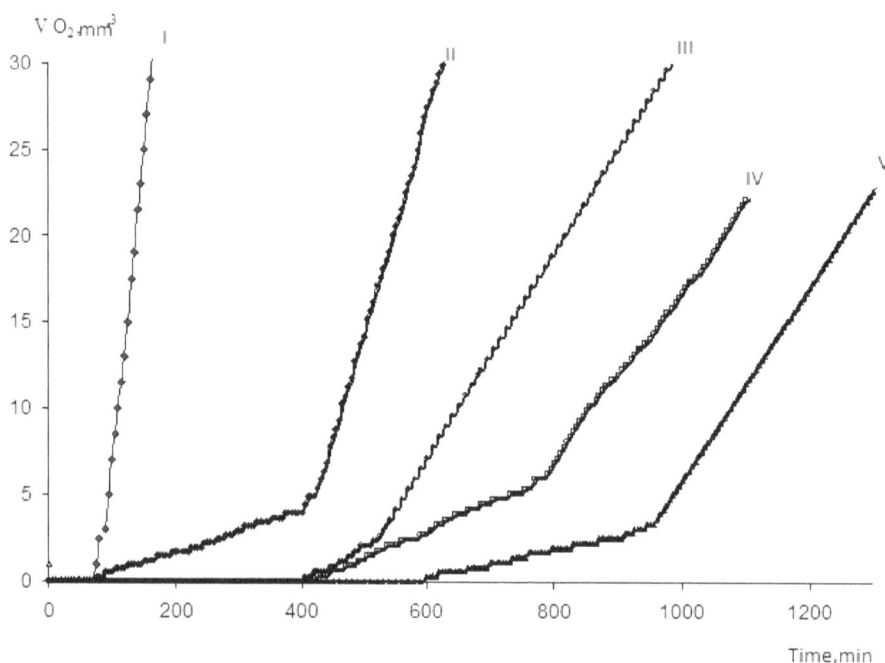

**Figure 9.** Kinetic curves of oxygen absorption of MO in the chlorobenzene medium upon initiation of the oxidation reaction by UV irradiation: I, substrate (S) (control); II, S + 3; III, S + 4; IV, S + 1; V, S + 2. Concentration of AO is $1.5 \bullet 10^{-4}$ mol L$^{-1}$, T = 60°C.

significantly reduces the effect of the antioxidant [31, 32, 34–36]. In the amide structure **4,** one of the two phenolic groups is spatially screened with bulky substituents, which provides a higher inhibitory effect 4 compared to **3**.

The kinetics of oxygen absorption in the oxidation of methyl oleate in the presence of amides 1–4 was studied in the range of concentrations $(0.5–2.0) \cdot 10^{-4}$ mol $L^{-1}$; the effectiveness of the antioxidants studied was compared with dibunol (**5**) and $\alpha$-tocopherol (**6**) (**Figure 9**). The linear character of the dependence of the period of induction on the concentration of amides, described earlier for most synthetic antioxidants, is established. For natural antioxidants, extreme dependence on concentration is observed; at relatively high concentrations, the inhibitory effect may decrease and a prooxidant effect may be observed [33]. It has been shown that the action of N-substituted amides of salicylic acids **1** and **2**, which are spatially hindered diatomaceous phenols, is comparable to the efficiency of monohydric phenol-5 and exceeds the gross inhibitory effect of a natural antioxidant **6**. Amide 1, which includes in its structure two phenolic groups, exceeds the monatomic phenol 5 in terms of the induction period by no more than 20%. Amides **3** and **4**, which differ in the highest degree of conjugation of the electron density, are significantly inferior to amides **1** and **2** and also amide **5** (**Figure 10**).

It is known that the chemical structure has a significant effect on the antioxidant properties of oxidation inhibitors. An important stage of the present study was the establishment of the relationship between the features of the chemical structure of a series of amides of salicylic acid and the mechanism of their antioxidant action.

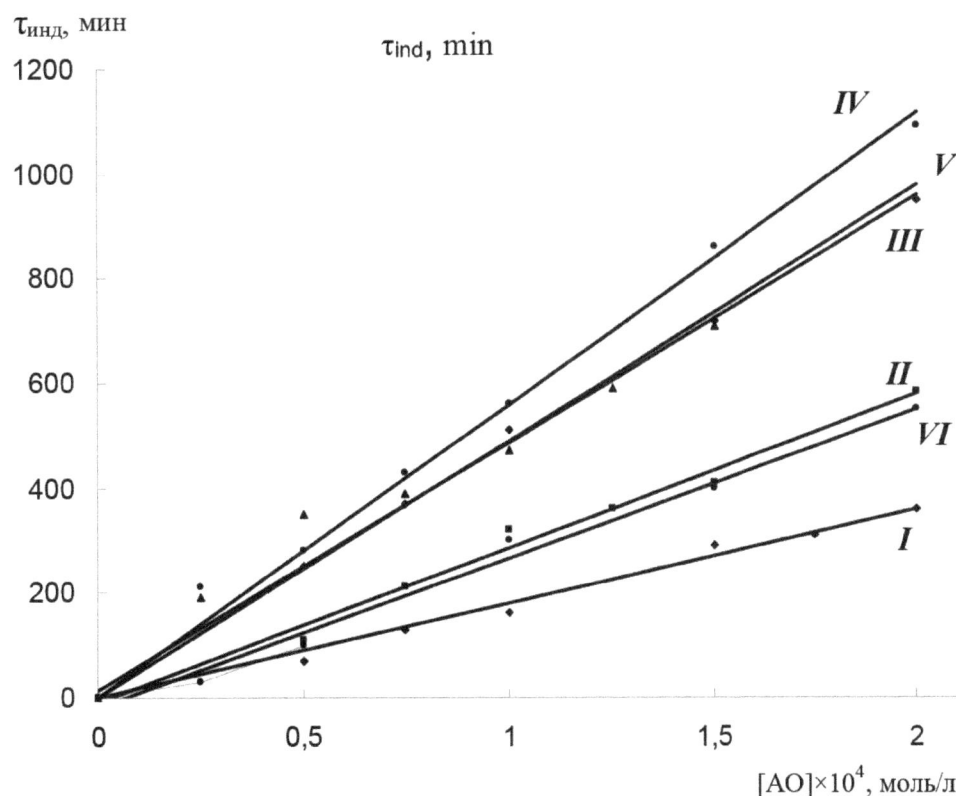

**Figure 10.** Dependence of the period of induction on the concentration of amides: I-3, II-4, III-2, IV-1, V-5, and VI-6 under UV-initiated oxidation of MO ($w_i = 0.6 \cdot 10^{-8}$ Mol $L^{-1} s^{-1}$).

The relatively low values of induction periods for N-substituted amides of salicylic acid (diatom phenols **1–4**) are found in comparison with monatomic phenols **5** and **6**. These differences are explained by the fact that compounds (**1–4**) form intra- and intermolecular hydrogen bonds (structure 2), in which some OH groups cannot participate in the reaction with peroxy radicals.

The effectiveness of amides of salicylic acid in the elemental reaction with peroxyl radicals $RO_2{}^\bullet$ is determined by the rate constant of the elementary reaction ($k_7$) (antiradical activity) estimated by the chemiluminescence method [12].

It has been established that all the derivatives of salicylic acid exhibit a high antiradical activity (**Table 1**).

**Table 1** shows that the highest antiradical activity is found in natural antioxidant **6** ($\alpha$-tocopherol), the efficacy of which is more than 400–460 times higher than that of compounds (**1–4**). The absence of bulky substituents in structure **5** provides this compound the highest antiradical activity. Note that all amides of salicylic acid have $k_7$ values of the same order as the known synthetic sterically hindered antioxidant (dibunol) (**5**). In the group of amides of salicylic acid, the highest values are characteristic for **3** and **4**, in the structure of which phenolic OH groups are included in the common conjugation system, which binds both benzene rings. For amides **1** and **2**, the conjugation system is "destroyed" by a bridging fragment containing three methylene groups, which contributes to a decrease in the value of $k_7$. The antiradical activity of amides of salicylic acid is comparable with the antioxidant activity of most known sulfur-containing and nitrogen-containing bifunctional oxidation inhibitors [36, 37].

It is known that the presence of phenolic OH groups with ortho-t-butyl substituents leads to a significant decrease in antiradical activity. Thus, the greatest value of $k_7$ was observed for compound **3**, which in its structure has two unsubstituted phenolic hydroxyls. The substituted analogue (amide **4**) is inferior to the **3** amide in **4**, 1 times, which is explained by the spatial difficulties in the interaction of antioxidant molecules with peroxy radicals. It is known that classical oxidation inhibitors interact, as a rule, with two free radicals (the stoichiometric inhibition coefficient for **5** and **6** is 2) [12, 36, 37]. The presence of two phenolic hydroxyls in N-substituted amides of salicylic acid theoretically should lead to an increase in the value of f to 4.0. However, the stoichiometric inhibition coefficient f approaches 4 only in two compounds (1 and 2) (f = 3.6–33), which kinetically reflects the presence of two OH groups in the molecule and indicates that both phenolic hydroxyls are interdependent during oxidation, and the difference in the value (f = 3.3–3.6) from 4.0 kinetically reflects the partial participation of

| The subject compound | 1 | 2 | 3 | 4 | 5 | 6 |
|---|---|---|---|---|---|---|
| $k_7 \times 10^4$, $M^{-1} \times c^{-1}$ | 0.85 | 0.52 | 6.86 | 1.69 | 1.40 | 360.0 |
| f | 3.6 | 3.3 | 2.4 | 2.6 | 2.0 | 2 |

T = 333 K, Wi = $2.3 \bullet 10^{-8}$, mol $L^{-1}$ $s^{-1}$, C = $1 \bullet 10^{-3}$ mol $L^{-1}$

**Table 1.** The values of the constant $k_7$ in the reaction of N-substituted amides of salicylic acid with peroxyl radicals ($RO_2{}^\bullet$).

phenolic groups in the formation of complexes with a hydrogen bond, which cannot participate in the reaction with free radicals. For amides 3 and 4, a high degree of $\pi$-$\pi$ and n-$\pi$ conjugation and a low degree of screening of the phenolic groups with tert-butyl substituents reduce the inhibition rate to (f = 2.4–2.6). It is known that spatially uncomplicated phenols easily penetrate into side reactions, which significantly reduces their overall antioxidant effect. One of the most significant adverse reactions is the interaction of phenoxy radicals with the substrate. Therefore, the weaker inhibitory ability of amides **3** and **4**, as well as the lower stoichiometric inhibition coefficient, indicates the occurrence of side reactions, in particular the possibility of the interaction of phenoxy radicals with the oxidation substrate.

# 6. Conclusions

1. Comparative analysis of the spectral absorption of native and partially deuterated molecules showed that amides of salicylic acid exist as complexes with an intramolecular hydrogen bond or an intermolecular hydrogen bond. Complexes are formed with the participation of phenolic and carbonyl groups located in neighboring positions. Spatially hindered phenolic groups, in all likelihood, do not participate in the formation of complexes.

2. During photolysis, N-substituted amides of salicylic acid pass into an excited state, which leads to the appearance of triplets and the formation of phenoxy radicals, presumably as a result of absorption of the second quantum of light by the excited singlet state. The triplet-triplet state is the main channel for the death of the triplet state of annihilation and recombination of phenoxy radicals ($k_{rek}$ = 2.3 · $10^8$ L $mol^{-1}$ $s^{-1}$). With UV irradiation of N-substituted amides of salicylic acid, amide groups also become excited; however, radical products are not formed in this case.

3. On the basis of a joint interpretation of pulsed and stationary photolysis data, stable photochemical conversion products of amides are established, which are compounds of the quinoid structure (iminoquinones). The presence of bulky substituents in the ortho- and para-positions prevents the recombination of phenoxy radicals with the formation of dimers. Iminoquinones are the final products of the photolysis reaction.

4. It is shown that all amides of salicylic acid are highly effective inhibitors of UV-induced oxidation of the model substrate (methyloleate). The inhibitory effect is directly proportional to the concentration of the compound, as for most known synthetic antioxidants.

5. Investigation of the relationship between the chemical structure and the antioxidant properties of a number of N-substituted amides of salicylic acid showed that the greatest inhibitory effect is manifested in compounds 1 and 2, in the structure of which the aromatic fragments are separated by three bridging groups –$CH_2$- and in ortho-position with respect to the phenolic group there are tert- butyl substituents.

6. It was found that the mechanism of photostabilizing action of N-substituted amides of salicylic acid is due to their ability to absorb ultraviolet rays in the range (190–350 nm), which creates prospects for using these compounds to protect the skin from melanoma

and to prevent the destruction of various materials under UV irradiation. It is shown that in the presence of amides of salicylic acid, the rate of initiation of the oxidation process is significantly reduced.

7. It is proven that the mechanism of action of N-substituted amides of salicylic acid is determined by their high activity in the reaction with free peroxy radicals. The value of the death constant for the radicals is ($k_7 = 0.52$–$6.86 \cdot 10^4$, L mol c$^{-1}$). The highest antiradical activity in amides 3 and 4 is noted, in the structure of which phenolic OH groups are included in the general conjugation system, covering both benzene rings. For amides 1 and 2, the conjugation system is "destroyed" by a bridging fragment containing three methylene groups, which contributes to a decrease in the value of $k_7$.

8. The inhibition factor f, showing the amount of free radicals dying on the antioxidant molecule, for compounds **1** and **2** approaches 4.0 (f = 3.6–3.3), reflecting the presence of two OH groups in the molecule, and indicates that both phenolic hydroxyls in the oxidation process act independently. Some of the phenolic groups are linked to complexes with a hydrogen bond and do not participate in the reaction with free peroxide radicals. Amides **3** and **4** show a high degree of $\pi$-$\pi$ and n-$\pi$ conjugation and a lower screening effect of bulky substituents. As a result, the inhibition coefficient decreases to (f = 2.4–2.6), and a significant role is played by side reactions, in particular the possibility of incorporating phenoxy radicals into the reaction with the substrate.

## Author details

Nadezhda Mikhailovna Storozhok* and Nadezhda Medyanik

*Address all correspondence to: nadinstor@mail.ru

Tyumen State Medical University, Ministry of Health of the Russian Federation, Odesskaya, Tyumen, Russian Federation

## References

[1]  Ludemann HC, Hillenkamp F, Redmond RW. The Journal of Physical Chemistry. A. 2000; **104**:3884

[2]  Kozma L, Khornyak I, Eroshtyak I, Nemet B. Zhurnal Prikladnoi Khimii. 1990;**53**:259 [J. Appl. Chem. USSR (Engl. Trcmsl.), 1990, **53**]

[3]  Storozhok NM, Medyanik NP, Krysin AP, Pozdnyakov IP, Krekov SA. Kinetika i Kataliz. 2012;**53**:170 [Kinet. Catal. (Engl. Transl), 2012, **53**]

[4]  Author's Certificate No. 1118012 USSR; Byul. Izobr. [Invention Bulletin], 1984 (in Russian)

[5]  Wang PY, Chen YP, Yang PZ. Dies and Pigments. 1996;**30**:141

[6] Pozdnyakov IP, Plyusnin VE, Grivin VP, Vorobyev DY, Bazhin NM, Vauthey E. Journal of Photochemistry and Photobiology, A: Chemistry. 2004;**162**:153

[7] Pozdnyakov IP, Sosedova YA, Plyusnin VF, Grivin VP, Bazhin NM. 2007;**55**:1270 [Russ. Chem. Bull. (Int. Ed.), 2007, **55**]

[8] Pozdnyakov IP, Plyusnin VF, Grivin VP, Vorobyev DY, Bazhin NM, Vauthey E. Journal of Photochemistry and Photobiology, A: Chemistry. 2006;**181**:37

[9] Mel'nikov MY, Ivanov VL. Eksperimentalhye Metody Khimicheskoi Kinetiki [Experimental Methods in Chemical Kinetics]. Fotokhimiya, Moscow. 2004. 125 pp. (in Russian)

[10] Storozhok NM, Medyanik NP, Krysin AP, Krekov SA, Borisenko VE. Zh. Org. Khim. 2013;**49**. 1046 [Russ. J. Org. Chem. (Engl. Transl), 2013, **49**]

[11] Denisov ET. Kinetics of Homogeneous Chemical Reactions. Moscow: Higher School; 1978. 367 p

[12] Emanuel NM, Denicov EN. Z.K. Mayzus Chain Reactions of Oxidation of Hydrocarbons in the Liquid Phase. Nauka: Moscow; 1966. p. 375

[13] Yongqing L, Yanzhen M, Yunfan Y, et al. Phisical Chemistry Chemical Phisics. 2018;**20**: 4208

[14] Rahangdale D, Kumar A, Anupama C, Archana C, Dhodapkar R. Journal of Molecular Recognition. 2017;**31**:S1 e 2630. http//doi 10.1002 /jmr.2630

[15] Khoa L,Kliaikin SY, Chulanovskii VM. in Molekulyarnaya spektroskopiya [Molecular Spectroscopy], 1973, Issue 2, 18 (in Russian)

[16] Rumynskaya IG, Shraiber VM. Molekulyarnaya spektroskopiya [Molecular Spectroscopy], Issue. 1986;**7**:132 (in Russian)

[17] Omura K. Journal of American Chemical Society. 1992;**69**:461

[18] Hiang H, Peng M, Li N, Peng S, Shi Hui J. Pharmaceutocal and Biomedical Analysis. 2017; **133**:75

[19] Shchepkin DN. Anharmonicheskie effekty v spektrakh kompleksov s vodorodnoi svyaz, yu [Anharmonic Effects in Spectra of Complexes with Hydrogen Bond]. 1987. deposited with YINITI. No. 7511-V. 87 (in Russian)

[20] Nagibina TI, Smolyanskii AL, Sheikh -Zade MI. Zhurnal Prikladnoi Khimii. 1982;**52**:754 [J. Organ. Chem. USSR (Engl. Transl), 1982, **52**]

[21] Bakker D, Ong Q, Arghya A, Jorome M, Marie-Pierre G, Anouk R. Journal of Molecular Spectroscopy. 2017:**342**:4

[22] Storohzok NM, Tsymbal IN, Petrenko NI, Schulz EE, Khrapova NG, Tolstikov GA, Burlakova EB. Biomedical Chemistry. 2003;**49**:96

[23] Kawski P, Kochel A, Perevozkina MG, Filarowski A. Journal of Molecular Structure. 2006; **790**:65

[24] Denisov GS, Sheikh-Zade MI, Eskina MV. Zhurnal Prikladnoĭ Khimii. 1977;**1049**:27 [J. Appl. Chem. USSR, 1977, **27**]

[25] Pikaev AK, Kabachi SA, Makarov IE, Ershov BG. Impul, snyi radioliz i ego primenenie [Pulse Radiolysis and Its Application]. Moscow: Atomizdat; 1980, 280 pp (in Russian)

[26] Sarakha M, Bolte M, Burrows HD. Journal of Photochemistry and Photobiology A: Chemistry. 1997;**107**:101

[27] Das TN. The Journal of Physical Chemistry. A. 2001;**105**:5954

[28] Hesse PJ, Battino R, Scharlin P, Wilhelm E. Journal of Chemical & Engineering Data. 1996; **41**:195

[29] Xibin G, Fangtong Z, Ralf IK. Chemical Physics Letters. 2007;**448**:7

[30] Hanway P, Jiadan J, Bhattacharjee U, et al. Journal of American Society. 2013;**135**:9078

[31] Shlyapintokh VY, Karpukhin ON, Postnikov LM. Chemiluminescent Methods for Studying Slow Chemical Processes. Moscow, Nauka; 1966. p. 138

[32] Roginskii VA. Fenoln, ye antioksidanty [Phenolic Antioxidants]. Nauka, Moscow; 1988. 247 pp. (in Russian)

[33] Eghbaliferiz S, Iranshahi M. Phytotheraphy Research. 2016;**30**:1379

[34] Zenkov NK, Lankin VZ, Menshchikova BV. Oxidative Stress. Moscow: Science / Interperiodica; 2001. p. 343

[35] Zenkov NK, Kanlantintseva VZ, NV Lankin, BV Menschikova, AE Prosenko. Phenolic Antioxidants. Novosibirsk: Siberian Branch of the Russian Academy of Medical Sciences; 2003. 328 p

[36] Storozhok NM, Gureeva NB, Khalitov RA, Storozhok AS, Krysin AP. Journal of Pharmaceutical Chemistry. 2011;**45**:732

[37] Storozhok NM, Perevozkina MG, Nikiforov GA. Russian Chemistry Bulletin. 2005;**54**:328 (Engl. Transl), 2005, **54**, 328

# 9

# Computational Study of the Photochemical Fragmentation of Hydantoin

Ming-Der Su

**Abstract**

The mechanism of the photochemical fragmentation reaction is investigated theoretically using the model system, hydantoin, using the CAS(22,16)/6-31G(d) and MP2-CAS-(22,16)/6-311G(d)//CAS(22,16)/6-31G(d) methods. The model investigation demonstrates that the preferred reaction route for the photofragmentation reaction is as follows: hydantoin → Franck-Condon region → conical intersection → fragment photoproducts (i.e., CO, isocyanic acid, and methylenimine). The theoretical finding additionally suggests that no organic radicals exist during the fragmentation reaction. Moreover, due to the high activation energy, the theoretical evidences suggest that it would be difficult to yield the three fragments under the thermal reaction. All the above theoretical observations are consistent with the available experimental results.

Keywords: hydantoin, 2,4-imidazolidinediones, photofragmentation, conical intersection, CASSCF

## 1. Introduction

Hydantoins (or 2,4-imidazolidinediones, **1**) are known to be useful chemicals in various pharmaceutical and agrochemical fields [1]. In order to examine the electronic structure, infrared spectrum, and unimolecular UV-induced photochemistry of the parent hydantoin monomer, Ildiz and co-workers recently reported that upon irradiation at $\lambda = 230$ nm, photochemical transformation of matrix-isolated **1** took place leading to formation of CO, isocyanic acid, and methylenimine (**Scheme 1**) [2]. However, since then neither experimental nor theoretical work has been devoted to the study of the photofragmentation mechanism of such five-membered ring heterocyclic molecule. In order to obtain more understanding of the photochemical

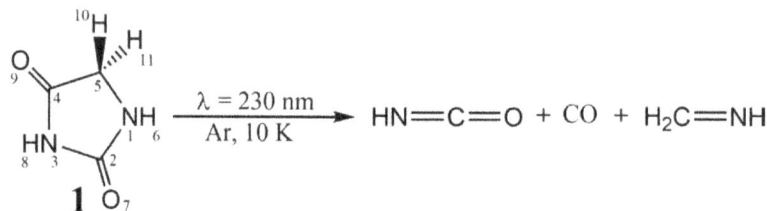

**Scheme 1.** The experimental result. See ref. [2].

behaviors of the transformation of hydantoin and their related heterocyclics to various photoproducts, the potential energy surfaces on its both singlet ground state and singlet excited state were investigated by CASSCF and MP2-CAS calculations. It will be shown below that the conical intersection (CI) [3–8], whose geometrical structure is fragments-like, plays a crucial role in the unimolecular photochemistry of hydantoin.

It is well established that most photochemical reactions begin from an excited potential surface and then cross over to a lower surface along the reaction route. They subsequently arrive on the ground state surface through a series of radiationless transitions (i.e., CIs). Eventually, they generate the photoproducts on the ground state surface [3–8]. That is to say, it is the presence of minima and transition states on the ground and excited state that controls the photochemical reactions [3–8].

# 2. Methodology

The theoretical results of the *ab initio* complete active space multiconfiguration self-consistent field (CASSCF) level of theory were achieved using the Gaussian 09 software package [9]. In this work, six $\sigma$, six $\pi$, and four nonbonding orbitals were selected as active orbitals. In addition, the optimization of CIs was achieved in the ($f$-2)-dimensional intersection space using the method of Robb et al. implemented in the Gaussian 09 program [9]. As a result, the 22 electrons in 16 orbitals CASSCF method was utilized with the 6-31G(d) basis sets (CAS(22,16)/6-31G(d)) for geometrical optimization.

The multireference Møller-Plesset (MP2-CAS) algorithm [10], which is given in the program package GAUSSIAN 09 [9], has also been utilized to compute dynamic electron correlations. In this work, the relative energies mentioned in the text are those determined at the MP2-CAS-(22,16)/6-311G(d) level using the CAS(22,16)/6-31G(d) (hereafter designed MP2-CAS and CASSCF, respectively) geometry.

# 3. Discussion

The central feature of the photochemical mechanism of **1** is the location of CI in the ground and excited electronic states. In this work, we shall use the molecular orbital (MO) model as presented in **Figure 1** as a basis for the interpretation of the phototransformation mechanism

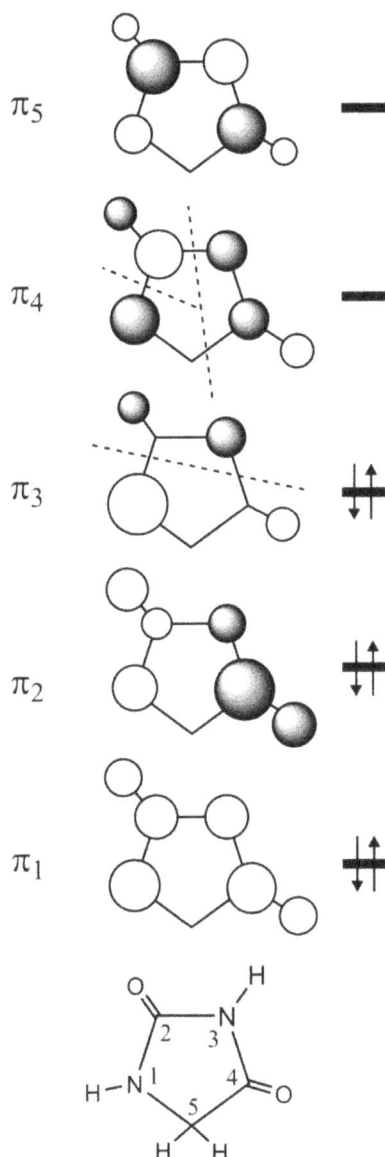

**Figure 1.** A $\pi$ molecular orbital diagram for hydantoin (**1**).

of **1**. From **Figure 1**, one may see that two node planes (dashed lines) existing on the lowest molecular orbital (LUMO) are between $N_1C_2$, $C_2N_3$, and $N_4C_5$ atoms, whereas only one node plane exists on the highest molecular orbital (HOMO). These phenomena strongly imply the photofragmentation behaviors of hydantoin (**1**).

**Figure 2** contains all the relative energies of the key points with respect to the energy of the reactant, **1**. Moreover, the geometrical structures of these points on the possible mechanistic pathway of **Figure 2** are demonstrated in the same figure.

At the beginning, **1** is promoted to its excited singlet state by a vertical excitation as shown in the left-hand side of **Figure 2**. Our MP2-CAS vertical excitation energy to the lowest excited $S_1$ state of **1** was computed to be 127 kcal/mol. However, according to the available experimental report

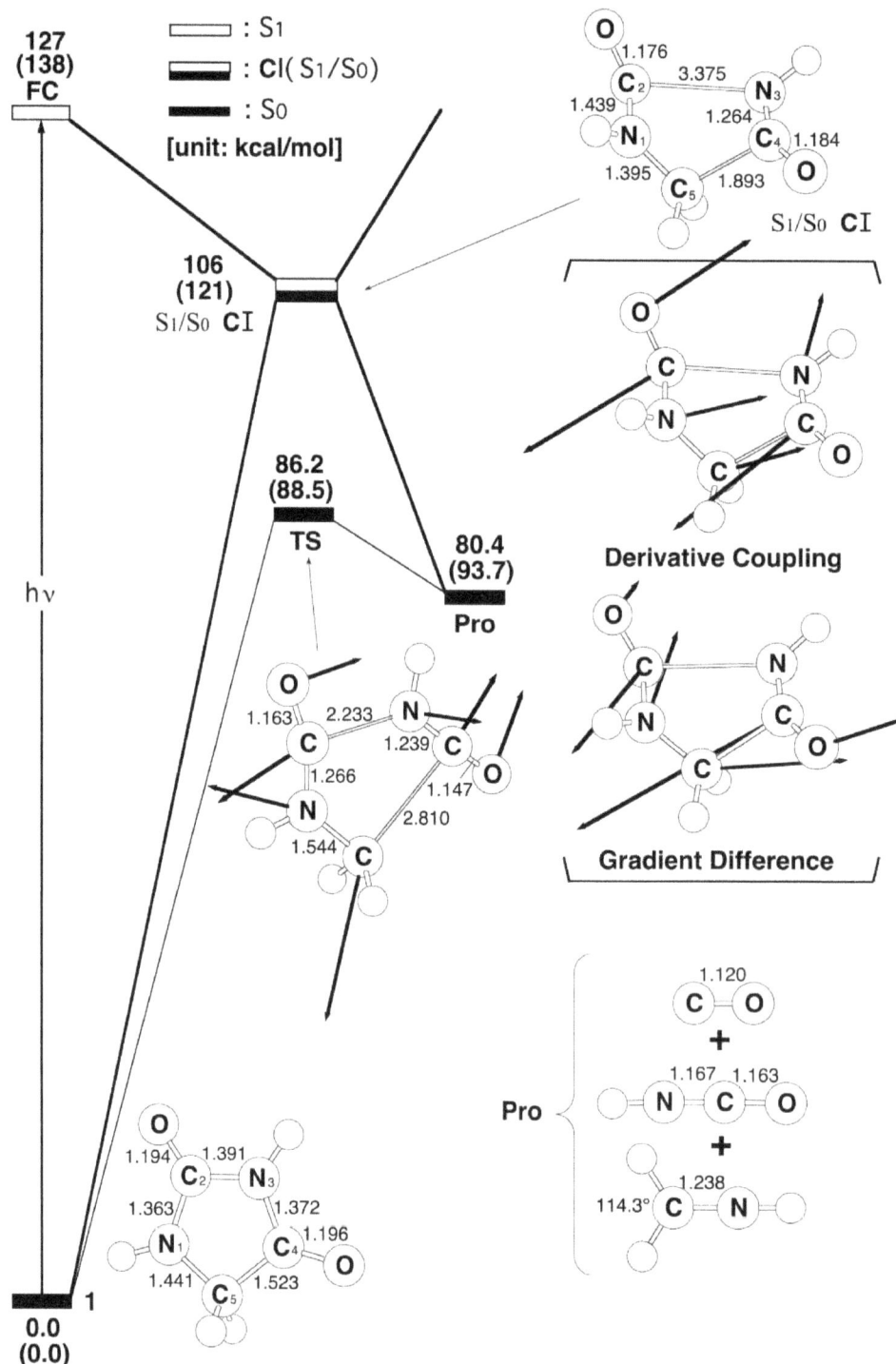

**Figure 2.** Potential energy surfaces for the fragmentation reaction of hydantoin (**1**). The abbreviations **FC**, **TS**, **CI**, and **Pro** stand for Franck-Condon, transition state, conical intersection, and products, respectively. The relative energies were obtained at the MP2-CAS-(22,16)/6-311G(d)//CAS(22,16)/6-31G(d) and CAS(22,16)/6-31G(d) (in parentheses) levels of theory. The selected geometrical parameters (Å) of CASSCF optimized structures of the stationary points are also given. Hydrogen is omitted for clarity. The heavy arrows in the **TS** indicate the main atomic motions in the transition state eigenvector. For more information, see the text.

shown in **Scheme 1**, the experimental wavelength $\lambda$ = 230 nm (=124.3 kcal/mol in energy) [2] is somewhat lower than our computational data. Nevertheless, it is believed that the present calculations should provide a good estimate of the relative energies for the parent hydantoin (**1**) system.

As seen in **Figure 2**, after the vertical excitation process, **1** is located on the excited singlet surface but still possesses the $S_0$ (ground state) geometry (**FC**). From the point reached by the vertical excitation, **1** relaxes to reach an $S_1/S_0$ **CI**, where the photoexcited system decays nonradiatively to $S_0$. Examining the geometrical structure of **CI** given in **Figure 2**, it is noteworthy that its $N_1$—$C_2$, $C_2$—$N_3$, and $C_4$—$C_5$ bond distances were calculated to be 1.439, 3.375, and 1.893 Å, respectively, which strongly imply that these chemical bonds are easily broken due to the node planes appeared between these atoms in the LUMO, as already schematically illustrated in **Figure 1**. The computational results predicted that the energy of $S_1/S_0$ **CI** lies 106 kcal/mol above **1** and 21 kcal/mol below **FC** at the MP2-CAS level of theory. Funneling through $S_1/S_0$ **CI**, different reaction pathways on the ground state surface may be anticipated by following the derivative coupling vector or the gradient difference vector directions.[2] **Figure 2** shows that the derivative coupling vector is mainly related to the $N_1C_2$, $C_2N_3$, and $C_4C_5$ stretching modes that give three kinds of fragment products (i.e., CO, isocyanic acid, and methylenimine) on the $S_0$ surface. On the other hand, gradient difference vector gives the asymmetric $C_2N_1C_5C_4$ bending motion that may lead to a vibrationally hot **1**-$S_0$ species. It has to be noted that this photoreaction path is a one-step process (the direct mechanism), which only involves the $S_1/S_0$ **CI** point. Besides, this work predicts that the reaction pathway for the photofragmentation of **1** should be a barrierless route. In consequence, the process of the photochemical reaction path can be represented as follows (**Figure 2**):

$$\mathbf{1} \rightarrow \mathbf{FC} \rightarrow \mathbf{S_1/S_0}\ \mathbf{CI} \rightarrow CO + H-N=C=O + H-N=CH_2$$

This work also examined the thermal reaction (**Scheme 1**) on the ground state ($S_0$) potential energy surface using the same levels of theory. In spite of the fact that photoexcitation elevates **1** into an excited electronic state, it was already emphasized that the products of the photochemical reaction are influenced by the $S_0$ potential surface [2]. If one searches for transition states on the $S_0$ surface near the structure of $S_1/S_0$ **CI**, one may obtain **TS**. From **Figure 2**, the energy of the **TS** connecting **1** and **Pro** on the $S_0$ surface lies about 20 kcal/mol below that of the $S_1/S_0$ **CI**. It should be mentioned here that the calculated results reveal the energy barriers for **1** → **Pro** and **Pro** → **1** are estimated to be 86 and 6.0 kcal/mol, respectively. This theoretical evidence indicates that it would be difficult to generate the three fragments (i.e., CO, isocyanic acid, and methylenimine) under the thermal (dark) reaction, which agrees well with the experimental observations [2].

Besides these, one may argue that it is possible to have radical formation in such a photofragmentation reaction. In fact, the CAS(22,16)/6-31G(d) method has been used to calculate the energies of some radicals (such as HN=C=O• and CO•). Their relative energies, however, are higher than that of the **CI** (121 kcal/mol, at most 133 kcal/mol). Therefore, the radical mechanism is not considered in this work.

## 4. Conclusion

In conclusion, the present theoretical computations demonstrate that upon absorption of a photon of light, the hydantoin (**1**) is excited vertically to $S_1$ (**FC**) via a $^1(\pi \rightarrow \pi^*)$ transition.

Subsequently, the excited **1** can enter an extremely efficient channel. From the **CI** point, **1** continues its development on the ground state potential surface to yield the photoproduct through either the radiationless path or return to **1** at the singlet ground state. Additionally, the theoretical examinations suggest that no other radicals exist in such a photofragmentation reaction. Also, the theoretical findings indicate that under the thermal procedures, the fragmentation reaction of **1** should be failed. All these theoretical findings can give successful explanations for the available experimental finding [2].

## Acknowledgements

The author is grateful to the National Center for High-Performance Computing of Taiwan for generous amounts of computing time, and the Ministry of Science and Technology of Taiwan for the financial support. The author also wishes to thank Professor Michael A. Robb, Dr. Michael J. Bearpark, Dr. S. Wilsey, (University of London, UK) and Professor Massimo Olivucci (Universita degli Studi di Siena, Italy) for their encouragement and support during his stay in London.

## Author details

Ming-Der Su[1,2]*

*Address all correspondence to: midesu@mail.ncyu.edu.tw

1 Department of Applied Chemistry, National Chiayi University, Chiayi, Taiwan

2 Department of Medicinal and Applied Chemistry, Kaohsiung Medical University, Kaohsiung, Taiwan

## References

[1] Opacic N, Barbaric M, Zorc B, Cetina M, Nagl A, Frkovic D, Kralj M, Pavelic K, Balzarini J, Andrei G, Snoeck R, Clercq ED, Raic-Malic S, Mintas M. The novel L- and D-amino acid derivatives of hydroxyurea and hydantoins: Synthesis, X-ray crystal structure study, and cytostatic and antiviral activity evaluations. Journal of Medicinal Chemistry. 2005;**48**:475-482

[2] Ildiz GO, Nunes CM, Fausto R. Matrix isolation infrared spectra and photochemistry of hydantoin. The Journal of Physical Chemistry. A. 2013;**117**:726-734

[3] Bernardi F, Olivucci M, Robb MA. Modelling photochemical reactivity of organic systems—A new challenge to quantum computational chemistry. Israel Journal of Chemistry. 1993:265-276

[4]  Bernardi F, Olivucci M, Robb MA. Potential energy surface crossings in organic photochemistry. Chemical Society Reviews. 1996;**25**:321-328

[5]  Klessinger M. Theoretical models for the selectivity of organic singlet and triplet photoreactions. Pure and Applied Chemistry. 1997;**69**:773-778

[6]  Klessinger M. Conical intersections and the mechanism of singlet photoreactions. Angewandte Chemie (International Edition in English). 1995;**34**:549-551

[7]  Klessinger M. Theoretical models for the selectivity of organic singlet and triplet photoreactions. Pure and Applied Chemistry. 1997;**69**,773-778

[8]  Klessinger M, Michl J. Excited States and Photochemistry of Organic Molecules. New York: VCH Publishers; 1995

[9]  Frisch MJ, Trucks GW, Schlegel HB, Scuseria GE, Robb MA, Cheeseman JR, Scalmani G, Barone V, Mennucci B, Petersson GA, et al. Gaussian, Inc., Wallingford CT; 2013

[10] McDouall JJW, Peasley K, Robb MA. A simple MCSCF perturbation theory: Orthogonal valence bond Møller-Plesset 2 (OVB MP2). Chemical Physics Letters. 1988;**148**:183-190

# 10

# Photochemistry of Lipofuscin and the Interplay of UVA and Visible Light in Skin Photosensitivity

Carolina Santacruz-Perez, Paulo Newton Tonolli,
Felipe Gustavo Ravagnani and Maurício S. Baptista

## Abstract

The topics about prevention against sunlight-induced damages and a secure threshold to light exposition have reached a bigger number of specialists in basic science and medical care. It has been accepted that ultraviolet light is very hazardous and visible light is safe, but recent studies from our group has shown that human keratinocytes exposed previously to ultraviolet A (UVA) light can generate an endogenous visible light-sensitive photosensitizer (lipofuscin), leading to higher levels of singlet oxygen, DNA damages and a wide-range of cellular insults due to intracellular lipofuscin accumulation. Disruption of cell death pathways and on essential metabolic processes, as autophagy and redox signaling, can collaborate to increase light-induced damages. We also discuss the importance of considering not only UVA but visible light too in protection against solar exposure as a way to prevent future pretumoral lesions.

**Keywords:** sun care, UVA radiation, visible light, lipofuscin, singlet oxygen, DNA damage

## 1. Introduction

The solar electromagnetic spectrum is composed of three main regions: ultraviolet (UV) (100–400 nm), visible (400–800 nm), and infrared (>800 nm). The UV range is subdivided into UVC (100–280 nm), UVB (280–315 nm), and UVA (315–400 nm), which show uneven ability to penetrate through the skin layers. UVC is broadly blocked by the ozone layer and has low penetration in epidermis rather than UVB and UVA. The UVB photons are directly absorbed by DNA, generating pyrimidine dimers which are well-known to promote carcinoma skin [1]. UVA and visible light are able to reach the deepest skin layers, as melanocytes layer. UVA has been thoroughly studied and implicated in carcinogenesis, especially, melanomas [2]. UVA and visible light are well-known to promote the oxidative stress by photosensitization reactions, generating reactive oxygen species, and oxidatively damaging biomolecules and organelles [3]. Among these biomolecules is the DNA, which exposed to photosensitization, can induce premutagenic lesions, as 8-oxo-dG, leading to mutagenesis if not repaired by DNA repair system [4]. Visible light has similar photosensitization mechanisms to UVA and high penetrability in the skin, but poor attention has been given to this light source in relation to skin carcinogenesis processes.

The harmful effects of exposure to ultraviolet (UV) radiation are widely known nowadays and it depends on the time of exposure (chronic or acute) [5]. The excessive exposure to ultraviolet radiation reflects much more serious effects: premature aging of the skin, immunosuppression, damages to the eyes, and it can be also associated to different types of skin cancer [6–8]. Our society is becoming aware that sun exposition to other wavelengths of light (not UV only) has important consequences to the skin health [9–11]. However, the molecular mechanisms involved are only starting to be uncovered.

The Scientific Committee on Emerging and Newly Identified Health Risks published an opinion on light sensitivity, which identified ultraviolet radiation as a risk factor for the aggravation of the light-sensitive symptoms in some patients bearing the chronic diseases actinic dermatitis and solar urticaria [12]. Moreover, scientific evidence relies on the potential impacts on public health caused by the artificial light, including the visible light spectrum, like in jaundice and other photosensitizing conditions [13, 14]. It is necessary to study further the impacts of visible and invisible ranges of light as well as characterizing some molecules that can act as photosensitizing agents and reduce side-effects of UV radiation from sunlight or from another light source. Additionally, several parts of the known visible spectrum can generate photosensitivity. However, the severity of the effect depends on the wavelength, intensity, and time of exposure.

Lipofuscin is a subproduct of cross-link reactions among oxidized lipids, proteins, and organelles, accumulated mainly in lysosomal compartment during oxidative stress [15]. Lysosomal hydrolases cannot digest lipofuscin that accumulates, progressively, inside the cells as electron-dense granules around the nucleus, especially during the post-mitotic period. Besides that, lipofuscin granules can incorporate transition metals, such as iron and copper, catalyzing Fenton or like-Fenton reactions, which generate the hydroxyl radical, a highly reactive oxygen specie that attacks DNA, protein, and lipids [16]. Many studies have investigated the properties and photochemistry of lipofuscin in retinal pigment epithelium (RPE) cells [17–21]. The lipofuscin of RPE cells has absorption peaks in the ultraviolet B range (280–330 nm) with emission in 570–605 nm [19, 20] and also absorbs blue light ($\lambda$ = 420 nm), generating singlet oxygen [21]. However, the cellular composition of lipids and proteins change among different cell types, assigning different photochemistry properties. Few studies have been done about lipofuscin in the human skin cells and its exact role in phototoxicity has never been described for that tissue.

In a recent publication by our group [22], we described that lipofuscin can act as an endogenous photosensitizer and lead to a higher sensitivity of human epidermal keratinocytes (HaCaT) to visible light. We observed that UVA dose of 12 J.cm$^{-2}$ stimulates autophagy inhibition and lipofuscin accumulation in keratinocytes, which promote singlet oxygen production and induce photodamage in cells when exposed to the visible light source. Here, we revised the main publications in this area and we show why lipofuscin indeed act as an additional photosensitizer to visible light in primary and immortalized human skin keratinocytes (NHK and HaCaT, respectively). This photosensitization of lipofuscin was able to increase the photooxidative processes inside the cell, generating singlet oxygen and promoting important damages in nucleic acids (FPG and Endo III in comet assays to identify DNA lesions).

## 1.1. UVA radiation followed by visible light causes higher cytotoxicity

Harmful effects caused by UVA radiation to eukaryotic cells have been thoroughly described [23]. Here, we quantified cell viability based on the reduction of 3 (4,5-dimethylthiazol-2-yl)-2,5-diphenyltetrazolium bromide (MTT) reagent, using different UVA radiation doses over immortalized human epidermal keratinocytes (HaCaT) and keratinocytes isolated from neonatal foreskin (NHK). Cells were exposed to increasing doses of UVA light and 48 h after irradiation we observed that HaCaT and NHK cell survival were around 50% at doses of 12 J. $cm^{-2}$ and 6 J.$cm^{-2}$, respectively (**Figure 1**). Depending on the light dose and cellular properties, such as the level of labile iron, cells will die mainly by necrosis and/or apoptosis [24]. However, live cells previously irradiated by UVA were then exposed to visible light and their survival rates were even lower than those ones obtained when we employed single light sources or inverted irradiation protocol (visible light – UVA) (**Figure 2**). This result led us to investigate whether UVA radiation could turn keratinocytes highly sensitive to visible light.

## 1.2. UVA radiation causes inhibition of autophagic flux and consequent lipofuscin accumulation

Profiles of apoptotic and necrotic cell death are very common in response to UVA exposure and they occur due to the prooxidant condition that severely decreases cell viability [25, 26].

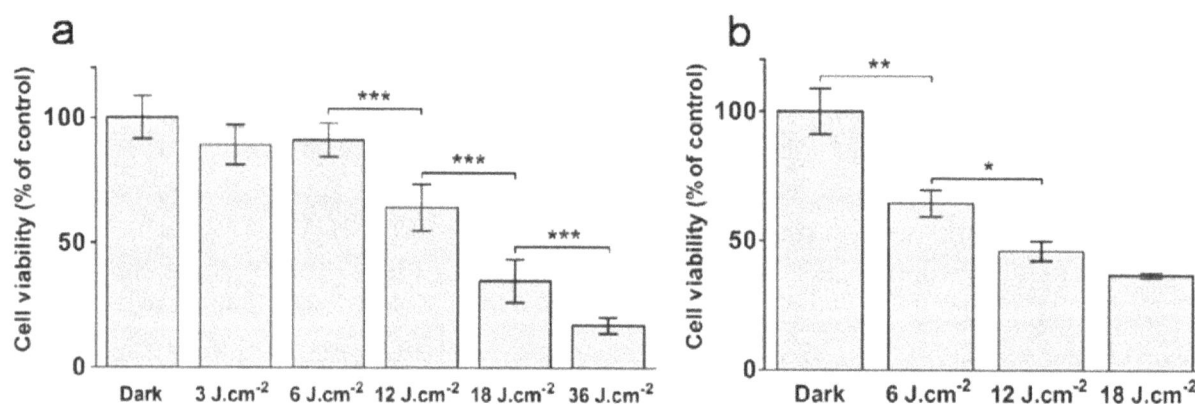

**Figure 1.** Viability curves as a function of UVA dose. Cell viability was measured 48 h after irradiation and is expressed as a percentage of MTT reduction by HaCaT (a) and NHK (b) cells. HaCaT and NHK cells reached survival of 50% at doses of 12 J.$cm^{-2}$ and 6 J.$cm^{-2}$, respectively. Bars indicate mean ± standard deviation (SD) for three independent experiments. Statistical analysis was performed using SigmaStat v.3.5 (Abacus Concepts, Berkeley, CA), performing one-way ANOVA test followed by Holm-Sidak posttest. Statistically significant differences are shown for $p < 0.05$ (*), $p < 0.01$ (**), and $p < 0.001$(***). From [22] with permission.

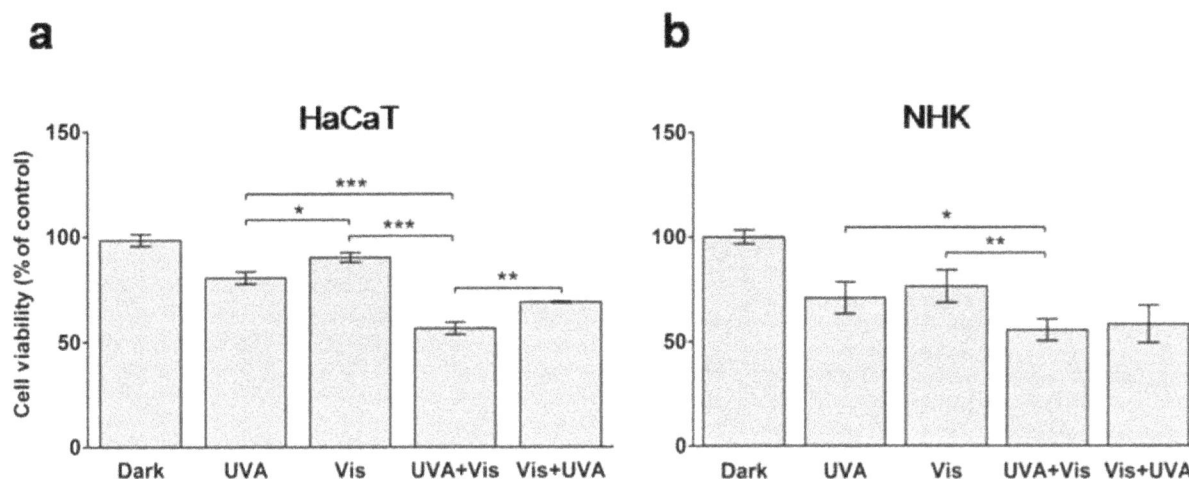

**Figure 2.** Viability of HaCaT (a) and NHK (b) cells after different light treatments based on MTT reduction (Visible = 36 J. cm$^{-2}$, UVA = 12 J.cm$^{-2}$). They were considered statistically significant differences for p-values <0.05 (*), <0.01(**) and < 0.001(***) in one-way ANOVA analysis. From [22] with permission.

There are few reports describing that redox misbalance and the formation of oxidized biomolecules can activate autophagy in normal and autophagy-deficient cells [27]. Lamore and co-workers showed that inactivation of lysosomal enzymes, such as cathepsins B and L, causes autophagic inhibition, with a consecutive accumulation of lipofuscin in dermal fibroblasts [28]. Therefore, UVA can lead to an efficient cell death concomitantly with autophagy or a sequential and temporal accumulation of lipofuscin in the survival cells, as previously suggested by Terman et al. [29] and Lamore and Wondrak [30].

We confirmed that lipofuscin accumulation in HaCaT cells after 48 h from exposure to UVA radiation at 12 J.cm$^{-2}$. It was possible to observe a five-fold higher values to the autofluorescence emission of lipofuscin in UVA-treated HaCaT. **Figure 3** shows this increment in both populations of cells (small and large ones). The scatter-plot contour of lipofuscin autofluorescence (FL1 Log) according to the cell size (FSC), obtained by flow cytometry in control cells and irradiated ones with UVA light. Average values of the population are represented in quadrant Q2 that characterize lipofuscin accumulation in UVA-treated cells (18 J.cm$^{-2}$) (**Figure 3b**) in relation to non-irradiated ones (dark) (**Figure 3a**).

The light absorption of lipofuscin in human skin keratinocytes extends from blue to green range of the visible spectrum. In fact, we detected a typical perinuclear fluorescence of lipofuscin (> 515 nm), when we excitate live cells with 450–490 nm. This fluorescence increased when primary and immortalized keratinocytes were exposed to growing doses of UVA, in a dose-dependent manner. Besides, lipofuscin-loaded keratinocytes showed a higher level of singlet oxygen generation when excited at 490 nm instead of another wavelengths (**Figure 8**). Therefore, lipofuscin granules from human skin keratinocytes seem more sensitive to blue light. Performing the time-resolved fluorescence microscopy, we detected the fluorescence lifetime (FLT) for lipofuscin granules, indicating a fluorescence emission lifetime around to 1.7 ns (**Figure 4**).

Lipofuscin FLT was already related as the fastest lifetime component of typically 190 ps in the retinal pigment epithelium [31]. Considering histograms of lifetimes after fluorescence excitation

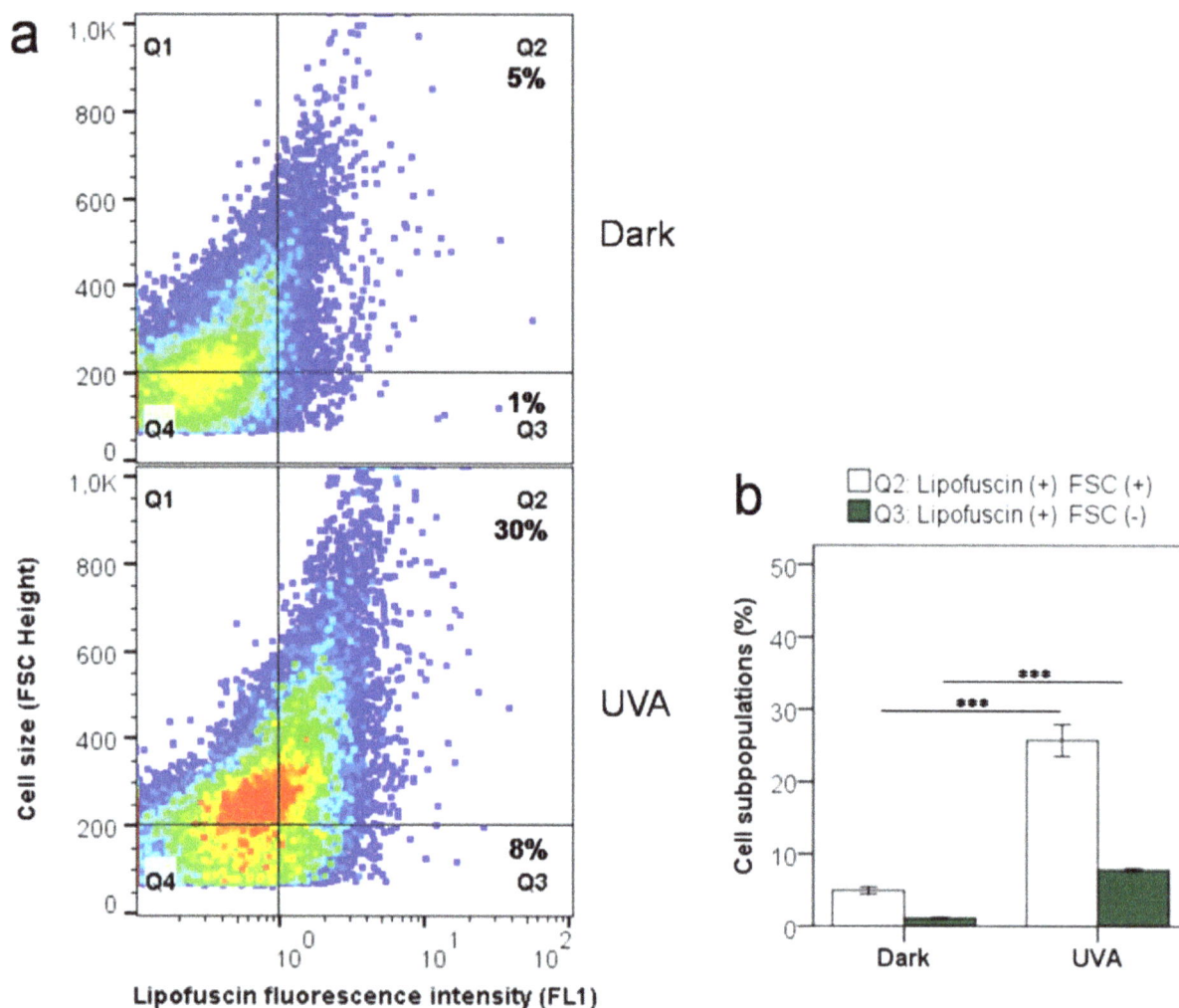

**Figure 3.** UVA radiation induces lipofuscin accumulation. (a) Scatter-plot contour of lipofuscin autofluorescence (FL1) versus forward-scattered light (FSC height), obtained by flow cytometry for dark control (upper diagram) and UVA-irradiated cells (18 J.cm$^{-2}$, lower diagram). (b) Mean values $\pm$ SD of cell subpopulations for dark control and UVA-irradiated represented in the quadrant Q2 {lipofuscin was accumulated [Lipofuscin (+)] and increased cell size [FSC (+)]}, and quadrant Q3 {lipofuscin was accumulated without increasing cell size [FSC ($-$)]}. Lipofuscin was excited at 488 nm and emission detected at 630 nm. Statistically significant differences were considered for p < 0.001(***). From reference [22] with permission.

and determined at the living human eye ground in the parapapillary region, a very low lifetime was calculated most frequently in the long-wavelength emission range ($\lambda$ > 500 nm) [32]. Our experiments showed that lipofuscin-loaded HaCaT cells (irradiated with UVA 12 J.cm$^{-2}$) can accumulate higher and perinuclear lipofuscin granules than those cells experimenting apoptosis (irradiated with 18 J.cm$^{-2}$, data not shown). Different lipofuscin composition can be conceived from FLT analysis since three lifetimes were obtained from cells irradiated at 6 and 12 J.cm$^{-2}$, and four lifetimes from irradiated cells at 18 J.cm$^{-2}$ of UVA light. A similar decrease of fluorescence lifetime has been observed for fixed and live cells [33]. No differences were found between FLT in artificially induced lipofuscin by another photosensitizer or by previous UVA light exposition (data not shown).

Accumulation of lipofuscin was also reported in dermal fibroblasts after UVA exposition [28]. We quantified lipofuscin fluorescence in HaCaT and NHK cells 48 h after the exposure to UVA

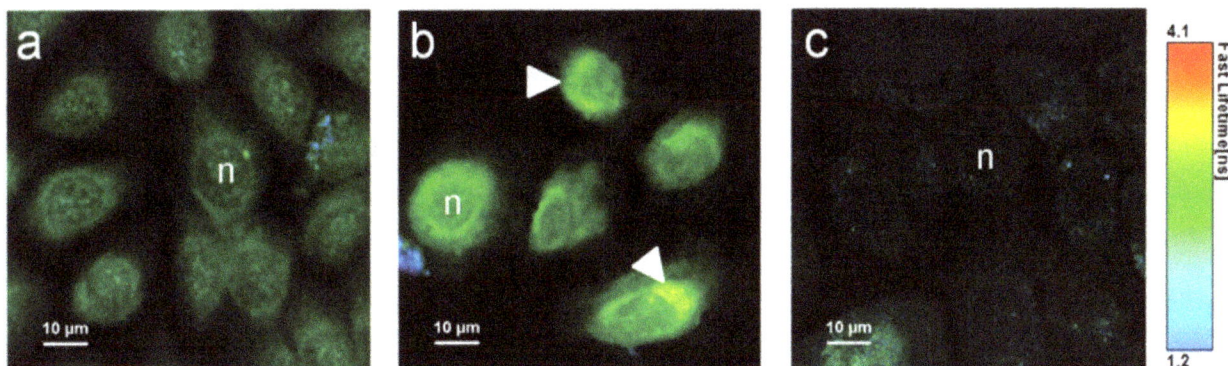

**Figure 4.** Lipofuscin accumulation and fluorescence lifetime in NHK cells. (a) Dark control. (b) NHK UVA-irradiated cells (6 $J.cm^{-2}$). White arrowheads indicate perinuclear aggregates of lipofuscin. (c) Visible light-irradiated cells (36 $J.cm^{-2}$). Images were obtained using a single molecule lifetime confocal microscope (Picoquant's Microtime 200). Samples were excited at 509 nm and emission was captured with a long-pass filter at 519 nm. Fluorescence decay curves of lipofuscin fluorophores were recorded by TCSPC mode. n = nucleus. From [22] with permission.

radiation (**Figure 5**), and we identified lipofuscin accumulation in the perinuclear region using Sudan Black B staining in both cell lineages (**Figure 6**).

Additionally, using transmission electron microscopy, we visualized electron-dense lipofuscin granules in the perinuclear region of HaCaT cells after 48 h from exposition to UVA radiation (**Figure 7**).

Although the accumulation of lipofuscin after UVA is evident, we needed further controls to definitively prove that it was lipofuscin photosensitization that caused an extra reduction in cell viability and not another mechanism induced by UVA. Note that UVA without formation of lipofuscin (inhibited by iron chelator) does not lead to a higher reduction in cell viability by visible light. The chemical induction of lipofuscin, employing a chemical

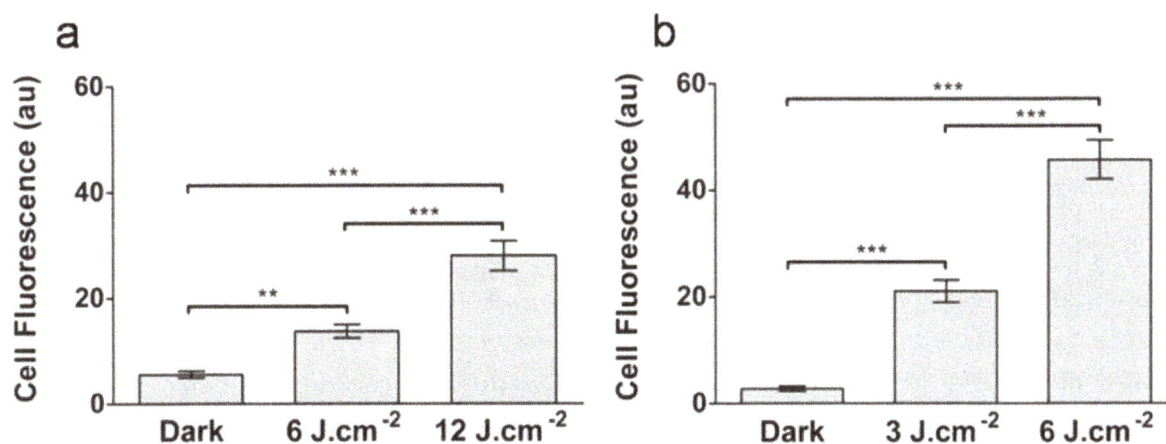

**Figure 5.** Lipofuscin fluorescence in HaCaT and NHK cells. Autofluorescence of HaCaT (a) and NHK (b) cells 48 h after exposure to UVA. Cell fluorescence was measured from TIF images, converted to 16-bit, and delimiting fluorescent area with polygonal selection, obtaining data per area, mean gray value, and integrated density from different images for each sample (HaCaT dark, 6 and 12 $J.cm^{-2}$; n = 11, n = 25, n = 20, respectively; NHK dark, 3 and 6 $J.cm^{-2}$; n = 5, n = 8, n = 8, respectively). Total cell fluorescence (TCF) was normalized using the following formula: TCF = integrated density − (selected area x mean fluorescence of background readings). Statistical analysis was performed using SigmaStat v.3.5. (Abacus Concepts, Berkeley, CA), performing one-way ANOVA test and Holm-Sidak posttest, considering statistically significant difference for $p < 0.01$ (**) and $p < 0.001$(***). From [22] with permission.

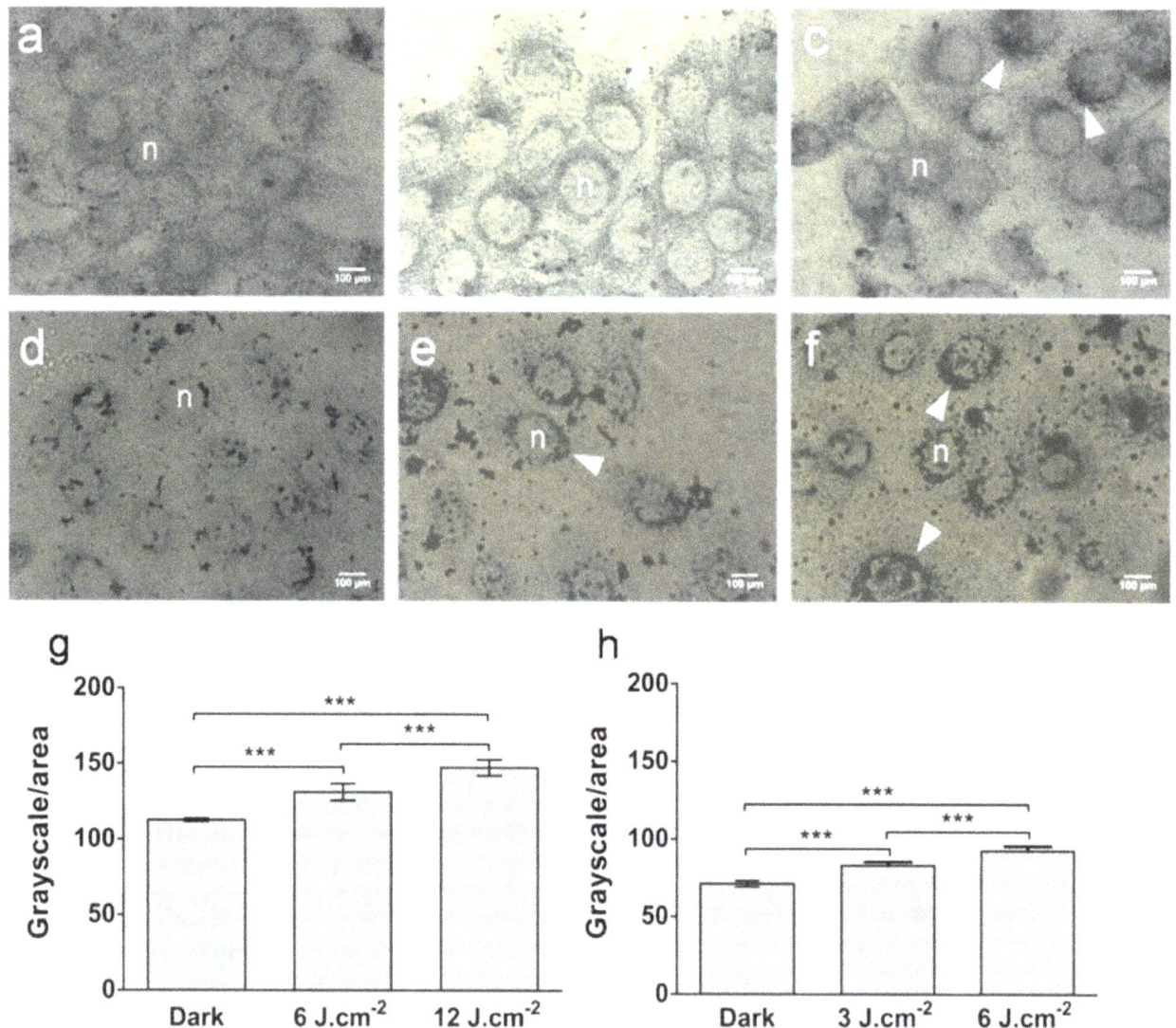

**Figure 6.** Lipofuscin identification by Sudan Black B staining in HaCaT and NHK cells. Both HaCaT and NHK cells were stained with Sudan Black B (SBB) 48 h after UVA radiation. Images were obtained using bright-field transmitted light microscopy. Whitehead arrows indicate lipofuscin stained by SBB. n = nucleus. For HaCaT cells are shown (a) Dark, (b) UVA 6 J.cm$^{-2}$, (c) 12 J.cm$^{-2}$ and for NHK cells (d) Dark, (e) 3 J.cm$^{-2}$ and (f) 6 J.cm$^{-2}$. Quantification of SBB staining in HaCaT (g) and NHK (h) cells. Images were converted to 8-bit grayscale. The sbb-stained area was delimited by adjusting cutoff levels of gray values in order to obtain area and mean gray value. Integrated intensity was divided by area from different images for each sample (HaCaT dark, 6 and 12 J.cm$^{-2}$; n = 3, n = 10, n = 10, respectively; NHK dark, 3 and 6 J.cm$^{-2}$; n = 6, n = 7, n = 4, respectively). One-way ANOVA analysis followed by Holm-Sidak posttest was performed in SigmaStat (v.3.5). Statistically significant differences are shown for p < 0.001 (***). From [22] with permission.

inhibitor of autophagy (chloroquine), causes this additional cell death, even without UVA irradiation (**Figure 8**).

## 1.3. Lipofuscin acts as a photosensitizer and causes pre-mutagenic lesions after the visible light exposition

Recently, we have shown that DNA of melanocytes suffers direct oxidative damage by melanin photosensitization with visible light [34]. We tested other skin lineages to verify if these damages can also occur from excessive exposition to visible light in response to the presence of

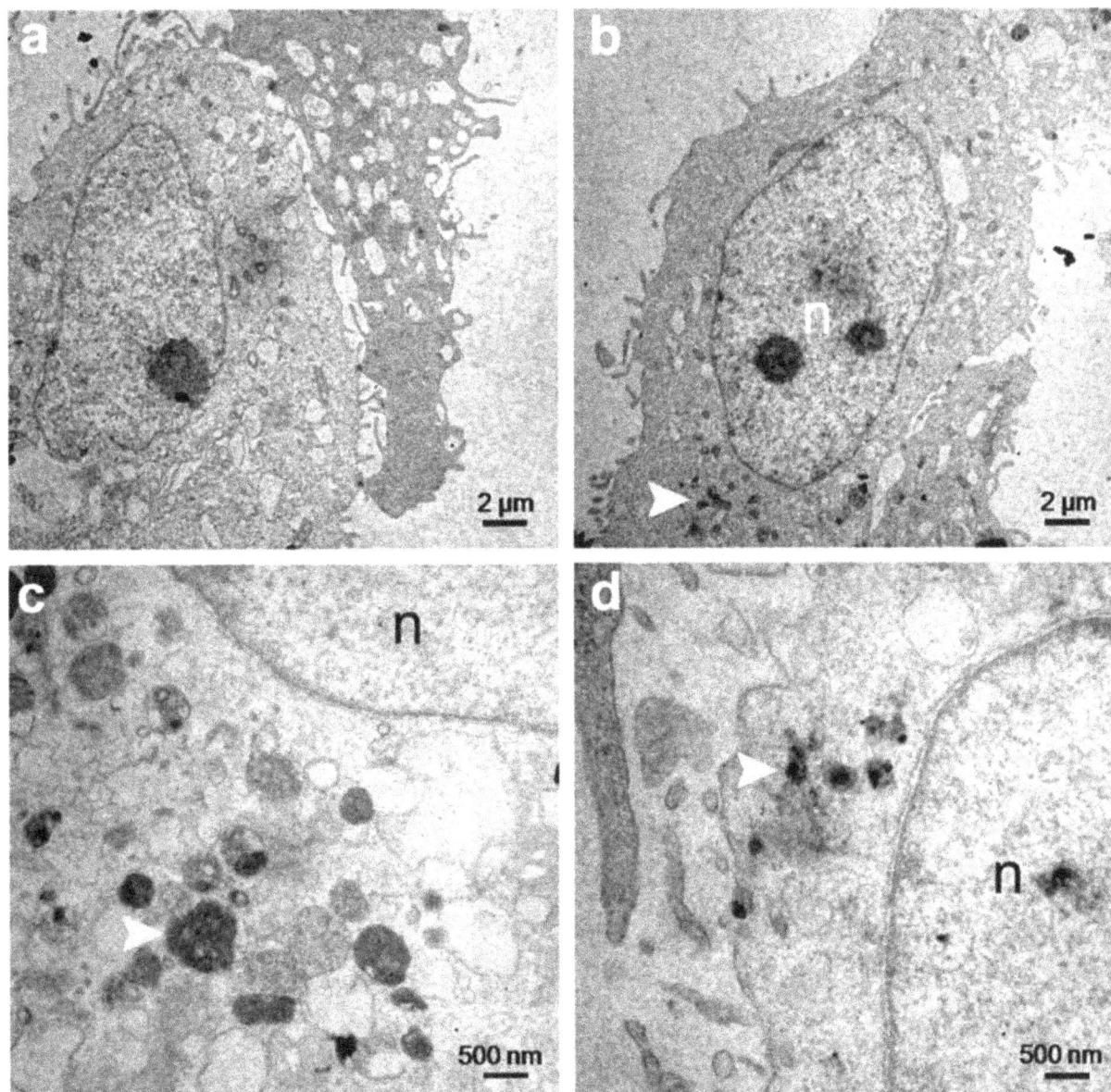

**Figure 7.** Lipofuscin accumulation in HaCaT cells after exposure to UVA by TEM. Lipofuscin was identified in the perinuclear region of HaCaT cells using transmission electron microscopy 48 h after exposition to UVA radiation. (a) Dark control. (b) UVA-irradiated cells ($6 \, J.cm^{-2}$) showing the typical accumulation of electron-dense granules of lipofuscin. (c) and (d) show lipofuscin in details. Whitehead arrows indicate lipofuscin granules. (n = nucleus). From [22] with permission.

natural photosensitizers as lipofuscin. To perform that, we calculated singlet oxygen production in HaCaT cells. For these cells, it was possible to identify that accumulated lipofuscin presented singlet oxygen phosphorescence emission after exposition to UVA light (data not shown) and increasing in DCF oxidation, indicating higher levels of chemical oxidizing species in UVA + Vis group (**Figure 9**). Moreover, if we treat previously HaCaT keratinocytes with a known UVA dose able to induce autophagy inhibition and lipofuscin accumulation (i.e. $12 \, J.cm^{-2}$), after sequential irradiation with visible light ($36 \, J.cm^{-2}$), it is only possible to observe significant reduction on cell viability and concomitant higher levels of DNA damage after the second irradiation step (increase in olive tail moment—OTM) (**Figures 10** and **11**). Control cells were irradiated with $36 \, J.cm^{-2}$ of visible light only, and they did not show phototoxicity.

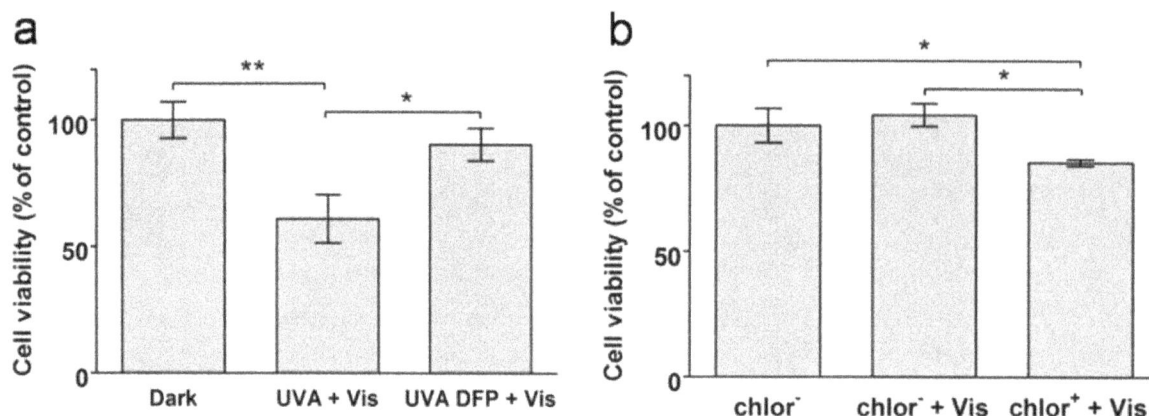

**Figure 8.** Cell viability based on the reduction of MTT in HaCaT cells. UVA radiation or by treatment with 60 μM chloroquine (a) UVA-treated cells (12 J.cm$^{-2}$) were immediately incubated with 30 μM deferiprone (DFP), an iron-chelator, avoiding lipofuscinogenesis. Forty-eight hours after initial challenge, cells were photosensitized with visible light (Vis) (36 J.cm$^{-2}$). (b) Cells incubated in the presence (chlor$^+$) or absence (chlor$^-$) of chloroquine (an autophagic inhibitor). Forty-eight hours after cells were photosensitized with visible light (Vis). MTT assay was measured after 48 h from the last irradiation. Bars indicate mean ± standard deviation. Statistical analysis was performed using SigmaStat v.3.5 (one-way ANOVA followed by Holm-Sidak posttest). Statistically significant differences are shown for $p < 0.05$ (*) and $p < 0.01$ (**). From [22] with permission.

Lipofuscin is a pigment that absorbs visible light and generates enough amounts of triplet and singlet oxygen [21]; therefore, it is likely that this pigment will engage in photosensitization reactions if cells were exposed to specific light sources, as sunlight. Interestingly, lipofuscin that has accumulated in keratinocytes-induced DNA damage, trigging type I and II photosensitization reactions when we employed visible light (see FPG and Endo III data in **Figure 10**).

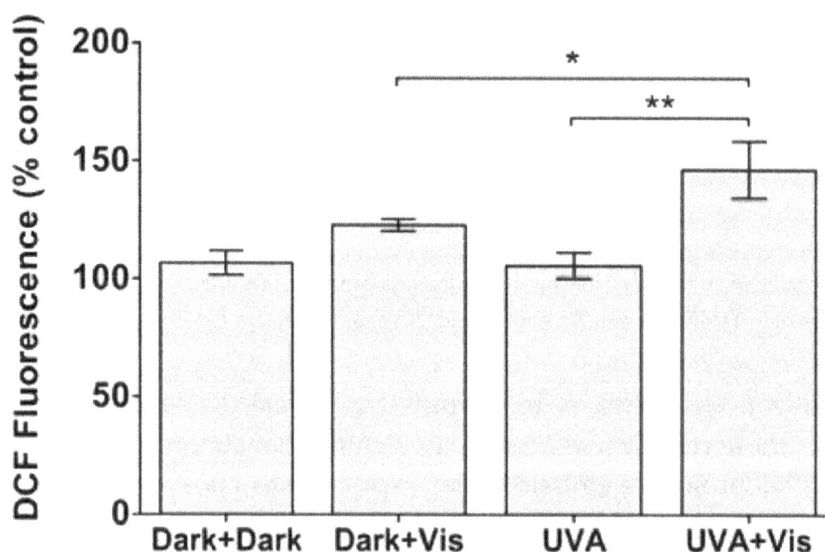

**Figure 9.** Dichlorofluorescein (DCF) assay for detection of oxidizing species in NHK cells measured immediately after the exposure to visible light. The increasing percentage of fluorescence per cell was calculated by $\%Fluorescence = [(Ft_{30} - Ft_0)/(Ft_0*100)]$, where $Ft_{30}$ is the fluorescence at 30 min and $Ft_0$ is fluorescence at 0 min. After 48 h from the first irradiation, cells were exposed to visible light (Dark+Vis and UVA + Vis). Bars indicate mean ± SD obtained from three independent cell culture experiments (n = 3). Statistical analysis was performed using SigmaStat v.3.5 (one-way ANOVA followed by Holm-Sidak posttest). Statistically significant differences are shown for $p < 0.05$ (*) and $p < 0.01$ (**). From [22] with permission.

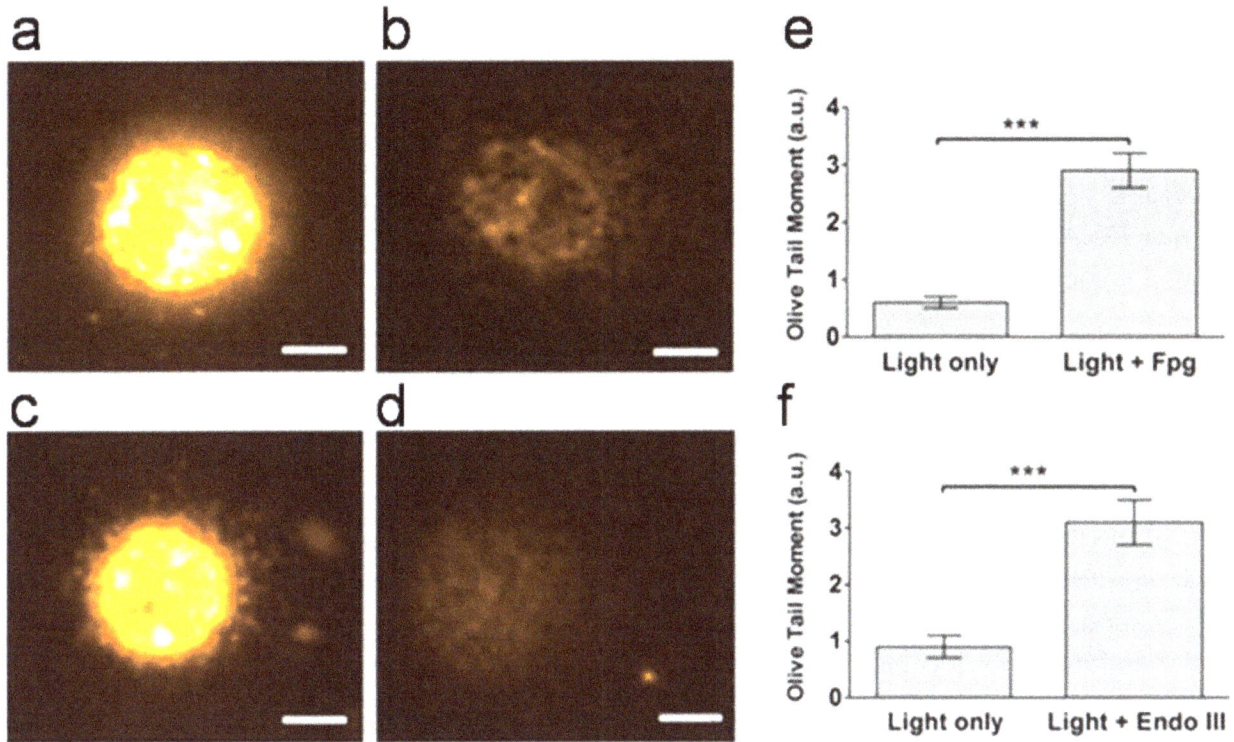

**Figure 10.** Visible light is phototoxic to lipofuscin-accumulating keratinocytes and causes oxidative DNA damage. Comet assays were performed after light treatment (UVA 6 J.cm$^{-2}$ and visible light 8 J.cm$^{-2}$) in the absence (a and c) and presence of FPG (b) and Endo III (d) enzymes. Quantification of olive tail moment (OTM) for (a) and (b) are shown in (e) and for (c) and (d) in (f). Statistically significant differences are shown for p < 0.001 (**). From [22] with permission.

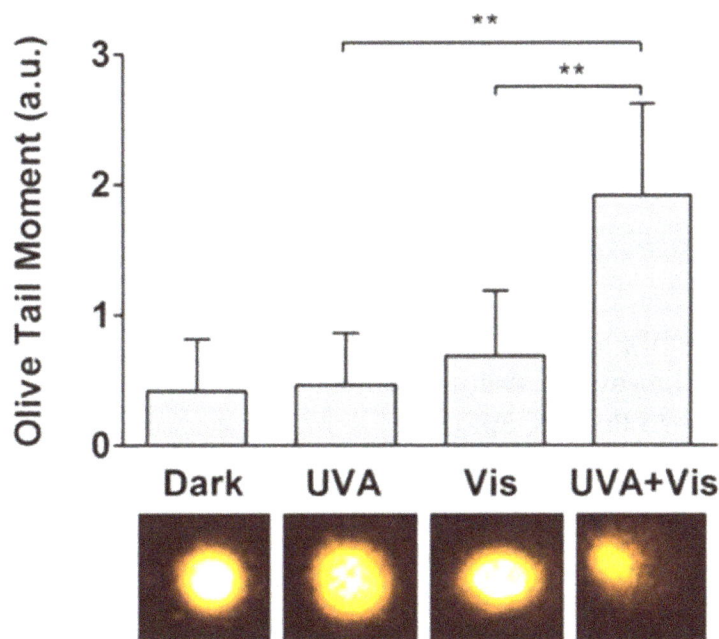

**Figure 11.** DNA fragmentation (comet assay) after irradiation protocols employing UVA and visible light sources. HaCaT cells were exposed to protocols with isolated light sources (UVA or visible light only) and a sequential protocol of irradiation (previous irradiation with UVA followed by visible light exposure). The graph shows the quantification of cumulative DNA damage when light sources are combined (see the figures right below the graph). Statistically significant differences were considered after one-way ANOVA analysis and Holm-Sidak posttest (**p < 0.01). From [22] with permission.

## 2. Conclusions

In this chapter, we reported that lipofuscin-loaded human skin keratinocytes presented higher sensitivity to visible light after their exposure to UVA radiation. Photosensitization of lipofuscin by visible light reduces cell viability, generating singlet oxygen and premutagenic lesions in nuclear DNA, as 8-oxo-dG. Here, we have indicated that lipofuscin can absorb the blue light more than other wavelengths, emitting fluorescence above 515 nm.

We consider that the exposure to risk factors like sunlight, which contain UVA and visible light, might have more attention as a public health problem and as a medical/dermatological alert for consumers, especially children. UVA radiation from not only sunlight can potentially increase the risk and incidence of skin cancer. UVA fingerprints are present in the majority of deeper-layer skin tumors and nowadays we have a similar situation occurring with the visible part of the light spectrum. The scientific community is clearly showing that this region of the solar spectra can have hazardous effects on the skin, but there are still few mechanistic explanations. Our results indicate that effects of UVA and visible light can amplify each other, and therefore, it is critical to start considering visible light in terms of human sun protection.

## Acknowledgements

This work was supported by FAPESP under grants 12/50680-5 and 13/07937-8, as well as, by NAP-Phototech and CNPq. We also thank all the co-authors of the original publication [22].

## Conflict of interest

The authors state no conflict of interest.

## Acronyms and abbreviations

Endonuclease-III (EndoIII); formamidopyrimidine [fapy]-DNA glycosylase (Fpg); 3(4,5-dimethylthiazol-2-yl)-2,5-diphenyltetrazolium bromide (MTT); human epidermal keratinocytes (HaCaT); Normal Human Primary Epidermal Keratinocytes isolated from Neonatal Foreskin (NHK).

## Author details

Carolina Santacruz-Perez, Paulo Newton Tonolli, Felipe Gustavo Ravagnani and Maurício S. Baptista*

*Address all correspondence to: baptista@iq.usp.br

Biochemistry Department, Institute of Chemistry, University of São Paulo, São Paulo, Brazil

# References

[1] Rochette PJ, Therrien JP, Drouin R, Perdiz D, Bastien N, Drobetsky EA, et al. UVA-induced cyclobutane pyrimidine dimers form predominantly at thymine-thymine dipyrimidines and correlate with the mutation spectrum in rodent cells. Nucleic Acids Research. 2003 Jun 1;**31**(11):2786-2794

[2] Wang SQ, Setlow R, Berwick M, Polsky D, Marghoob AA, Kopf AW, et al. Ultraviolet A and melanoma: A review. Journal of the American Academy of Dermatology. 2001 May;**44**(5):837-846

[3] Kielbassa C, Roza L, Epe B. Wavelength dependence of oxidative DNA damage induced by UV and visible light. Carcinogenesis. 1997 Apr;**18**(4):811-816

[4] Ghosh R, Mitchell DL. Effect of oxidative DNA damage in promoter elements on transcription factor binding. Nucleic Acids Research. 1999 Aug 1;**27**(15):3213-3218

[5] Doubova SV, Infante-Castaneda C. Factors associated with quality of life of caregivers of Mexican cancer patients. Quality of Life Research. 2016 Nov;**25**(11):2931-2940

[6] Ransohoff KJ, Jaju PD, Tang JY, Carbone M, Leachman S, Sarin KY. Familial skin cancer syndromes: Increased melanoma risk. Journal of the American Academy of Dermatology. 2016 Mar;**74**(3):423-434 quiz 35-6

[7] Wu S, Han J, Laden F, Qureshi AA. Long-term ultraviolet flux, other potential risk factors, and skin cancer risk: a cohort study. Cancer Epidemiology, Biomarkers & Prevention. 2016 Jun;**23**(6):1080-1089

[8] He H, Wisner P, Yang G, Hu HM, Haley D, Miller W, et al. Combined IL-21 and low-dose IL-2 therapy induces anti-tumor immunity and long-term curative effects in a murine melanoma tumor model. Journal of Translational Medicine. 2006 Jun 13;**4**:24

[9] Schieke SM, Schroeder P, Krutmann J. Cutaneous effects of infrared radiation: from clinical observations to molecular response mechanisms. Photodermatology, Photoimmunology & Photomedicine. 2003 Oct;**19**(5):228-234

[10] Mahmoud BH, Ruvolo E, Hexsel CL, Liu Y, Owen MR, Kollias N, et al. Impact of long-wavelength UVA and visible light on melanocompetent skin. The Journal of Investigative Dermatology. 2010 Aug;**130**(8):2092-2097

[11] Chiarelli-Neto O, Ferreira AS, Martins WK, Pavani C, Severino D, Faiao-Flores F, et al. Melanin photosensitization and the effect of visible light on epithelial cells. PLoS One. 2014;**9**(11):e113266

[12] Ahlbom A, Bridges J, de Seze R, Hillert L, Juutilainen J, Mattsson MO, et al. Possible effects of electromagnetic fields (EMF) on human health–opinion of the scientific committee on emerging and newly identified health risks (SCENIHR). Toxicology. 2008 Apr 18;**246**(2–3):248-250

[13] Kaae J, Boyd HA, Hansen AV, Wulf HC, Wohlfahrt J, Melbye M. Photosensitizing medi-

cation use and risk of skin cancer. Cancer Epidemiology, Biomarkers & Prevention. 2010; **19**(11):2942-2949

[14]  O'Gorman SM, Murphy GM. Photosensitizing medications and photocarcinogenesis. Photodermatology, Photoimmunology & Photomedicine. 2014;**30**(1):8-14

[15]  Jung T, Bader N, Grune T. Lipofuscin: Formation, distribution, and metabolic consequences. Annals of the New York Academy of Sciences. 2007 Nov;**1119**:97-111

[16]  Hohn A, Jung T, Grimm S, Grune T. Lipofuscin-bound iron is a major intracellular source of oxidants: role in senescent cells. Free Radical Biology & Medicine. 2010 Apr 15;**48**(8): 1100-1108

[17]  Boulton M, Docchio F, Dayhaw-Barker P, Ramponi R, Cubeddu R. Age-related changes in the morphology, absorption and fluorescence of melanosomes and lipofuscin granules of the retinal pigment epithelium. Vision Research. 1990;**30**(9):1291-1303

[18]  Reszka K, Eldred GE, Wang RH, Chignell C, Dillon J. The photochemistry of human retinal lipofuscin as studied by EPR. Photochemistry and Photobiology. 1995 Dec;**62**(6): 1005-1008

[19]  Eldred GE, Katz ML. Fluorophores of the human retinal pigment epithelium: separation and spectral characterization. Experimental Eye Research. 1988 Jul;**47**(1):71-86

[20]  Eldred GE, Miller GV, Stark WS, Feeney-Burns L. Lipofuscin: Resolution of discrepant fluorescence data. Science. 1982 May 14;**216**(4547):757-759

[21]  Rozanowska M, Wessels J, Boulton M, Burke JM, Rodgers MA, Truscott TG, et al. Blue light-induced singlet oxygen generation by retinal lipofuscin in non-polar media. Free Radical Biology & Medicine. 1998 May;**24**(7–8):1107-1112

[22]  Tonolli PN, Chiarelli-Neto O, Santacruz-Perez C, Junqueira HC, Watanabe IS, Ravagnani FG, et al. Lipofuscin generated by UVA turns keratinocytes photosensitive to visible light. The Journal of Investigative Dermatology. 2017 Nov;**137**(11):2447-2450

[23]  Ziegler A, Jonason AS, Leffell DJ, Simon JA, Sharma HW, Kimmelman J, et al. Sunburn and p53 in the onset of skin cancer. Nature. 1994 Dec 22-29;**372**(6508):773-776

[24]  Zhong JL, Yiakouvaki A, Holley P, Tyrrell RM, Pourzand C. Susceptibility of skin cells to UVA-induced necrotic cell death reflects the intracellular level of labile iron. The Journal of Investigative Dermatology. 2004 Oct;**123**(4):771-780

[25]  Moan J, Peak MJ. Effects of UV radiation of cells. Journal of Photochemistry and Photobiology. B. 1989 Oct;**4**(1):21-34

[26]  Tyrrell RM, Keyse SM. New trends in photobiology. The interaction of UVA radiation with cultured cells. Journal of Photochemistry and Photobiology. B. 1990 Mar;**4**(4):349-361

[27]  Zhao Y, Zhang CF, Rossiter H, Eckhart L, Konig U, Karner S, et al. Autophagy is induced by UVA and promotes removal of oxidized phospholipids and protein aggregates in epidermal keratinocytes. The Journal of Investigative Dermatology. 2013 Jun;**133**(6):1629-1637

[28] Lamore SD, Wondrak GT. UVA causes dual inactivation of cathepsin B and L underlying lysosomal dysfunction in human dermal fibroblasts. Journal of Photochemistry and Photobiology. B. 2013 Jun 5;**123**:1-12

[29] Terman A, Kurz T, Navratil M, Arriaga EA, Brunk UT. Mitochondrial turnover and aging of long-lived postmitotic cells: the mitochondrial-lysosomal axis theory of aging. Antioxidants & Redox Signaling. 2010 Apr;**12**(4):503-535

[30] Lamore SD, Wondrak GT. Autophagic-lysosomal dysregulation downstream of cathepsin B inactivation in human skin fibroblasts exposed to UVA. Photochemical & Photobiological Sciences. 2012 Jan;**11**(1):163-172

[31] Schweitzer D, Schenke S, Hammer M, Schweitzer F, Jentsch S, Birckner E, et al. Towards metabolic mapping of the human retina. Microscopy Research and Technique. 2007 May;**70**(5):410-419

[32] Schweitzer D, Hammer M, Schweitzer F. Limits of the confocal laser-scanning technique in measurements of time-resolved autofluorescence of the ocular fundus. Biomedizinische Technik. Biomedical Engineering. 2005 Sep;**50**(9):263-267

[33] Nardo L, Kristensen S, Tonnesen HH, Hogset A, Lilletvedt M. Solubilization of the photosensitizers TPCS(2a) and TPPS(2a) in aqueous media evaluated by time-resolved fluorescence analysis. Die Pharmazie. 2012 Jul;**67**(7):598-600

[34] Chiarelli-Neto O, Pavani C, Ferreira AS, Uchoa AF, Severino D, Baptista MS. Generation and suppression of singlet oxygen in hair by photosensitization of melanin. Free Radical Biology & Medicine. 2011 Sep 15;**51**(6):1195-1202

# 11

# Photochemical Degradation of Organic Xenobiotics in Natural Waters

Sarka Klementova

**Abstract**

Xenobiotics in the environment include a wide variety of compounds, e.g. pesticides, drugs, textile dyes, personal care products, stabilisers, and many others. These compounds enter natural waters by rain washing of treated areas, via leaching through soil from places of application and via waste waters of manufacturing facilities or municipal waste waters (excretion of unmetabolised drugs, disposal of unused drugs). In natural waters, physical, chemical, and biological processes contribute to the decrease of xenobiotics concentrations. For substances resistant to biological degradation processes and the chemical reactions such as hydrolysis, photoinitiated processes may represent important degradation pathways. Photochemical processes can be categorised in connection with the environmental fate of xenobiotics into two fundamental groups: those that may occur in natural waters and those that have been tested for decontamination of waste waters. The first group is focused mainly on photosensitization and homogeneous photocatalysis. The second class comprises advanced oxidation processes (AOPs) of which especially heterogeneous photocatalysis on semiconductors is the most investigated technique. The chapter covers all these processes and brings examples of their applications.

**Keywords:** natural waters, xenobiotics, emerging pollutants, photochemical degradation, advanced oxidation processes

## 1. Introduction

While most water assessments emphasise water quantity, water quality is also critical to satisfying basic human and environmental needs. The quality of the world's water is under increasing threat as a result of population growth, expanding industrial and agricultural activities, and climate change. Poor water quality threatens human and ecosystem health, increases water treatment costs, and reduces the availability of safe water for drinking and other uses [1]. It also limits economic productivity and development opportunities. Indeed, the United Nations find that "water quality is a global concern as risks of degradation translate directly into social and economic impacts" [2].

Human society relies on rivers for many functions and services including drinking water, irrigation, navigation, transport, recreation, and waste disposal. It has been estimated that despite accounting for just 0.4% of the Earth's surface area and 0.006% of the Earth's freshwater, rivers contain 6% of all described species and provide 5.1% of global ecosystem services [3, 4].

Water quality concerns are widespread, though the true extent of the problem remains undisclosed. In developing countries, an estimated 90% of sewage and 70% of industrial waste are discharged into waterways entirely untreated [5].

In recent times, anthropogenic activities, namely the production and consumption of chemically manufactured products, have been linked to growing environmental pollution and resulting health challenges. Currently, the pollution of the global water cycle with persistent organic contaminants appears to be one of the most important challenges of the twenty-first century. The majority of these organic substances are only partially removed by conventional wastewater treatment plants; hence they enter the environment and spread across different ecological compartments.

Most of the persistent contaminants are unregulated or in the process of regulation, yet they possess possible toxic effects in long-term exposure or potent endocrine disrupting properties, which lie in their interference with the hormonal function of living organisms including humans [6, 7].

Therefore, the studies of many scientists are focused on getting information on the environmental occurrence of the compounds belonging to this vast family of species. Other studies investigate the environmental fate of these compounds as well as the feasibility of their degradation in wastewaters. Among these studies, those concentrating on the photochemical processes and techniques have recently acquired a particular attention.

## 2. Organic micropollutants in the environment

Chemicals of emerging concern have no clearly stated definition; therefore, no comprehensive list of them exists. Kümmerer [8] defined emerging micropollutants as unregulated compounds or those with limited regulation which are present in the environment at low range (µg/l and below), irrespective of their chemical structure, and which thus require monitoring. Marcoux et al. [9] summarised emerging micropollutants as newly detected substances in the environment or those already identified as risky and the use of which in manufactured items is prohibited, or substances already known but the recent use of which in products may cause problems during their future treatment as waste. According to the US Geological Society [10], emerging contaminants are any synthetic or naturally occurring chemical or any microorganism or metabolite that is not commonly monitored in the environment but has the potential to

| Family/use | Emerging contaminant | Prescription 2012 (kg) | Surface waters (ng/l) |
|---|---|---|---|
| Antibacterial | Amoxicillin | 158,231 | 2.5–245 |
| | Erythromycin | 41,057 | 0.5–159 |
| | Metronidazole | 12,300 | 1.5–12 |
| | Ofloxacin | 219 | — |
| | Oxytetracycline | 17,143 | — |
| | Trimethoprim | 10,998 | 1.5–108 |
| Non-steroid anti-inflammatory drug | Paracetamol (acetaminophen) | >2,000,000 | 1.5–1388 |
| | Ibuprofen | 108,435 | 1–2370 |
| | Naproxen | 126,258 | 1–59 |
| | Ketoprofen | 243 | 1–4 |
| Lipid regulator | Simvastatin | 49,198 | <0.6 |
| | Bezafibrate | 7966 | 10–60 |
| Beta blocker | Propranolol | 9076 | 0.5–107 |
| | Atenolol | 20,725 | 1–487 |
| | Metoprolol | 2311 | 0.5–10 |
| Calcium channel blocker | Diltiazem | 21,922 | 1–17 |
| Hypertension | Valsartan | 6484 | 1–55 |
| Antidepressant | Venlafaxine | 16,211 | 1.1–35 |
| | Amitriptyline | 10,171 | <0.6–30 |
| | Fluoxetine | 5319 | 5.8–14 |
| | Dosulepin | 3270 | 0.5–25 |
| | Nortriptyline | 439 | 0.8–6.8 |
| Antiepileptic | Gabapentin | 104,110 | 0.6–1879 |
| Hypnotic | Temazepam | 883 | 3.2–34 |
| | Diazepam | 335 | 0.6–0.9 |
| | Oxazepam | 85 | 2.4–11 |
| Sunscreen agent | 1-benzophenone | | 0.3–9 |
| | 2-benzophenone | | 0.5–18 |
| | 3-benzophenone | | 15–36 |
| | 4-benzophenone | | 3–227 |
| Preservative | Methylparaben | | 0.3–68 |
| | Ethylparaben | | 1–13 |
| | Propylparaben | | 0.2–7 |
| | Butylparaben | | 0.3–6 |

**Table 1.** Examples of emerging contaminant occurrence for wastewaters and surface waters in the United Kingdom—based on the review by Petrie et al. [12].

enter the environment and cause known or suspected adverse ecological and/or human health effects.

An overview of micropollutants, their sources and effects, and their occurrence in different types of water including analytical detection techniques and concentration ranges is provided by an outstanding review by Tijany et al. [11]. Another comprehensive review by Petrie et al. [12] presents information about contaminants occurrence in wastewaters and surface waters in the United Kingdom, spatial distribution and seasonality, possibilities of microbial transformation, and possible ecotoxicological effects, together with some recommendations for the environmental monitoring of these substances. As can be seen from the lists of the substances covered in the above-mentioned studies as emerging xenobiotic compounds, categories such as pharmaceuticals (antibiotics, antidiabetics, antiepileptics, anti-inflammatories, analgetics, antidepressants) and personal care products (disinfectants, preservatives) are the most abundant representatives. Selected substances from the study of Petrie et al. [12] which represent the highest load on the environment are presented in **Table 1**. According to recent studies, more the 200 different pharmaceuticals alone have been reported in river waters globally to date, with concentrations mainly in the ng/l to μg/l range in surface waters [13–17], but in some cases, concentrations of even several orders of magnitude higher have been reported. Effluent concentrations from pharmaceutical formulation facilities in the USA (New York) reached 1.7 mg/l for the analgesic oxacodone and 3.8 mg/l for the muscle relaxant metaxalone [18]. Li et al. [19] found very high concentrations of tetracycline derivatives up to 800 and 2 mg/l in the effluent and receiving waters, respectively, from a sewage treatment plant serving an antibiotic manufacturing facility in China.

Besides these emerging contaminants, there are other organic pollutants, e.g. pesticides, chiefly herbicides, the presence of which in the aquatic environment has been known for a long time [20–23].

Most of the xenobiotics detected in natural waters are persistent compounds that are recalcitrant to microbial decay and resist chemical degradation through hydrolysis or other chemical reactions. Many of them contain aromatic rings, heteroatoms, and functional groups that can either absorb solar radiation or react with photogenerated transient species in natural waters (e.g. reactive oxygen species and/or photoexcited natural organic matter). Some of these compounds carry functional groups and structures such as phenol, carboxyl, nitro, and naphthyloxy that have been found to undergo photodegradation [24]; many of the pesticide compounds are chlorine derivatives, which predispose them to dechlorination and hydroxyderivative formation [25].

## 3. Photoinitiated reactions related to the organic xenobiotics degradation

Each reaction started by absorption of radiation may be classified as a photochemical or photoinitiated reaction. According to the mechanism of the photoinitiated reaction related to the degradation of xenobiotics, photolytic, photosensitized, and photocatalytic reactions can be distinguished.

### 3.1. Photolytic reactions

A photolytic reaction is usually understood as a reaction in which the absorbed quantum of radiation has enough energy to cause the breaking of a covalent bond in the substrate compound. Usually, highly energetic UV radiation (254 nm) is necessary for this purpose. The reaction includes only one reactant, the molecule that undergoes photolysis; therefore, the reaction follows first-order kinetics.

### 3.2. Photosensitised reactions

A photosensitised reaction needs a sensitizer molecule. This is a molecule that can absorb radiation and transfer the absorbed excitation energy onto another molecule. The energy can be transferred either onto an organic molecule, substrate (xenobiotic compound), or onto an oxygen molecule, which results in the formation of singlet oxygen. The possible reactions are illustrated in Eqs. (1)–(5).

$$^1Sens + h\nu \rightarrow \; ^1Sens^* \tag{1}$$

$$^1Sens^* + \; ^1Substrate \rightarrow \; ^1Substrate^* + \; ^1Sens \rightarrow Product + \; ^1Sens \tag{2}$$

$$^1Sens^* \rightarrow through \; ISC \rightarrow \; ^3Sens^* \tag{3}$$

$$^3Sens^* + \; ^3O_2 \rightarrow \; ^1O_2 \tag{4}$$

$$^1O_2 + \; ^1Substrate \rightarrow Oxidised \; product \tag{5}$$

Eq. (1) represents excitation of the sensitizer from the ground state (which is always a singlet state, i.e. all electrons in the molecule are paired) to the first excited singlet state. Eq. (2) represents energy transfer onto the substrate and its subsequent reaction into a product. Eq. (3) shows the possible conversion of the sensitizer from the first excited singlet state into the first triplet state (where two electrons are unpaired) through so-called intersystem crossing (ISC). The sensitizer in the triplet state is able to react with molecular oxygen dissolved in the reaction mixture (Eq. (4)) because the ground state of molecular oxygen with its two unpaired electrons is a triplet state. The reaction provides an excited form of oxygen, singlet oxygen, which is a powerful oxidative species; singlet oxygen then can react with organic substrate molecules and oxidise them (Eq. (5)).

Humic substances are considered to be the most common naturally occurring sensitisers.

Humic substances, comprising two major classes, humic acids and fulvic acids, are organic constituents of not only soil humus and peat but also streams, dystrophic lakes, and ocean water. They are produced by the biodegradation of dead organic matter as products of microbial metabolism although they are not synthesised as a life-sustaining carbon structures or compounds serving as energy storage. A typical humic substance is not a single, well-defined molecule, but a mixture of many molecules which typically include aromatic nuclei with carboxylic and phenolic groups as demonstrated in the structure proposed by Stevenson [26],

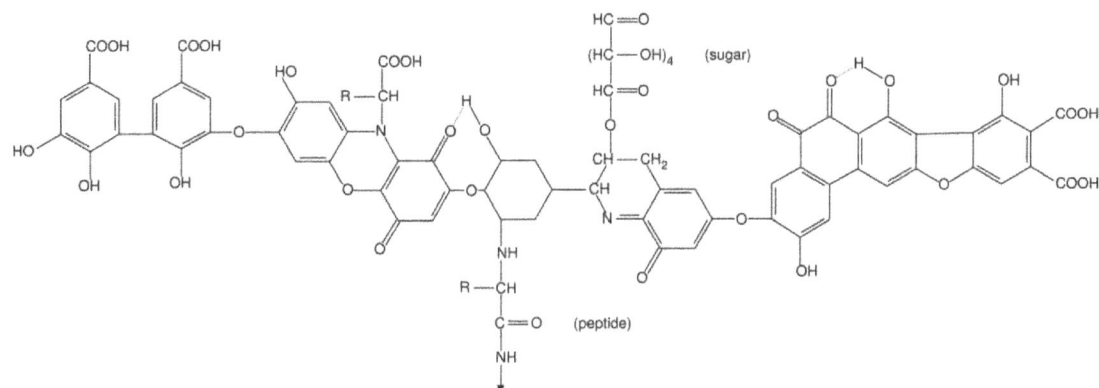

**Figure 1.** Proposed structure of humic acids [26].

which is illustrated in **Figure 1**. Their molecular weight ranges from a few hundred to several million of daltons [27].

The distinction between humic and fulvic acid is based on their solubility: humic acids are soluble in water at neutral and alkaline pH values and insoluble at acid pH, while fulvic acid is soluble in water across the full range of pH. Fulvic acids have usually smaller molecules and less extent of aromaticity, which results in less content of phenolic groups and more hydroxylic groups in side chains.

These functional groups contribute most to the surface charge and reactivity of humic substances. Humic and fulvic acids behave as mixtures of dibasic acids with a $pK_1$ value of around 4 for protonation of carboxyl groups and around 8 for protonation of phenolate groups [28].

In natural waters, myriads of other sensitisers can be found—natural pigments such as heme/porphyrine-based molecules (chlorophylls, bilirubin, hemocyanin, haemoglobin), carotenoids, or flavonoids (anthocyanins), but all of these are present in extremely low concentrations in the water environment and are therefore not considered to be of real significance for photochemical transformation of organic xenobiotic compounds.

### 3.3. Photocatalytic reactions

Photocatalysis may occur as a homogeneous process or as a heterogeneous process.

In homogeneous photocatalytic reactions, light contributes to the production of a catalytically active form of a catalyst. One example of such a reaction is the photochemically induced reduction of ferric ions in the presence of an electron donor to ferrous ions that exhibit much higher catalytic activity in comparison with the oxidised form [29, 30]. The subsequent catalytic reaction of a substrate is a 'dark' reaction, i.e. not photochemical, since the reaction does not need light. The active form of the catalyst enables the otherwise spin-forbidden reaction between a singlet substrate and triplet dissolved molecular oxygen.

Homogeneous photocatalytic reactions also include the so-called photo-Fenton reactions. Fenton's reagent is a solution of hydrogen peroxide with ferrous ions as a catalyst of an oxidative reaction with organic substrates. The reagent was described by H. J. H. Fenton in

1894. The sequence of reactions leading to the formation of reactive oxygen species (hydroxyl radicals and superoxide radicals) is represented in Eqs. (6)–(9).

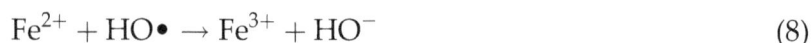

$$Fe^{2+} + H_2O_2 \rightarrow HO\bullet + Fe^{3+} + HO^- \tag{6}$$

$$Fe^{3+} + H_2O_2 \rightarrow HOO\bullet + Fe^{2+} + H^+ \tag{7}$$

$$Fe^{2+} + HO\bullet \rightarrow Fe^{3+} + HO^- \tag{8}$$

Since ferrous ions are quickly reoxidised in a Fenton reaction mixture, the photochemical variant is used; in the photo-Fenton reaction, the ferric ions are photochemically reduced in situ. Additional sources of OH radicals through photolysis of $H_2O_2$ and through the reduction of $Fe^{3+}$ ions under UV light (Eqs. (9) and (10)) are thus obtained.

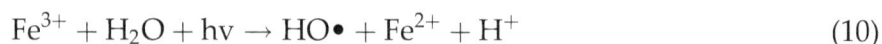

$$H_2O_2 + h\nu \rightarrow 2\,HO\bullet \tag{9}$$

$$Fe^{3+} + H_2O + h\nu \rightarrow HO\bullet + Fe^{2+} + H^+ \tag{10}$$

Heterogeneous photocatalysis is usually understood as any photochemical reaction on a semiconductor.

Semiconductor photocatalysis uses solid catalytic systems while the substrate to be degraded is dissolved or dissipated in the solution (or in the gaseous phase) around the catalyst. Five distinct steps in the process of the reaction on a semiconductor are involved:

- The transfer of liquid or gaseous phase reactant to the catalytic surface by diffusion.

- The adsorption of the reactant on the catalyst surface.

- The reaction of the adsorbed molecules.

- The desorption of products.

- The removal of products from the interface region by diffusion.

The initiation of the photocatalytic process involves the photochemical formation of an electron-hole pair after the absorption of a photon of sufficient energy for the excitation of an electron from the valence band of the semiconductor to its conduction band. The holes and electrons react with the solvent (water) and dissolved oxygen to produce oxidative species, mainly OH and superoxide radicals by the sequence of reactions presented in Eqs. (11)–(16).

$$h^+ + H_2O \rightarrow HO^\bullet + H^+ \tag{11}$$

$$h^+ + OH^- \rightarrow HO^\bullet \tag{12}$$

$$O_2 + e^- \rightarrow O_2^{\bullet-} \tag{13}$$

$$O_2^{\bullet-} + H^+ \rightarrow HO_2^\bullet \tag{14}$$

$$2HO_2^\bullet \rightarrow H_2O_2 + O_2 \tag{15}$$

$$H_2O_2 + O_2^{\bullet-} \rightarrow HO^\bullet + O_2 + OH^- \tag{16}$$

Various metal oxides were tested in semiconductor photocatalytic reactions. The most frequently used is not only $TiO_2$ [31–33] but also other materials such as ZnO [34, 35], ZnS [36], $Fe_2O_3$ [37], and many others including semiconductor composites as well as semiconductors doped with precious metals or modified semiconductors [31, 38–41].

Heterogeneous photocatalytic reactions are usually described by first-order kinetics though actually pseudo-first order kinetics (with OH radicals in excess to the reactant) is the relevant kinetic model.

# 4. Photochemistry of organic xenobiotics

## 4.1. Environmental relevance of the study approaches

Generally, three types of photochemical studies of xenobiotic compounds can be recognised:

- Environmental photochemical studies, i.e. field or laboratory studies under natural conditions (sun irradiation or irradiation with the source simulating as much as possible the solar spectrum, concentrations of substances the same as in natural aquatic systems).

- Studies relevant to environmental conditions (irradiation sources simulating part of the solar spectrum, concentrations of organic substrates usually higher than those found in natural waters).

- Other photochemical studies, using short wavelengths not relevant to environmental conditions, addition of chemicals such as $O_3$, $H_2O_2$, additional components such as semiconductors acting as photocatalysts.

The first two approaches enable us to quantify transformation rates, identify photoproducts, and estimate the importance of the photochemical reactions to the mass balance of pollutants in the natural environment or at least (in the second case) make an extrapolation to the environmental conditions relatively straightforward.

The third approach is not aimed at the elucidation of environmental processes; its goal lies in the removal of polluting substances remaining in wastewaters after the application of the classical procedure consisting of microfiltration, ultrafiltration, reverse osmosis, activated carbon absorption, and sand filtration [42]. To achieve this goal, techniques involving advanced oxidation processes (AOPs) are usually applied [43, 44].

For environmental photochemical studies or for those relevant to the environmental condition, solar spectrum radiation must be considered. UV region represents only a small part of the solar spectrum; it is estimated that the region from 290 to 380 nm forms ca 5% of the overall ultraviolet plus visible range [45]. Although UV solar radiation reaching the Earth's surface represents a small part of the solar spectrum, it plays an important role in its photochemical effects since it contains the radiation of highest energies. Nevertheless, considering photochemical reactions of xenobiotics, it is necessary to keep in mind that these compounds often exhibit negligible or none absorption in the region of wavelengths longer than 290 nm, since their absorption lies in the shorter UV wavelength range; therefore, many of the compounds cannot react directly under sun irradiation. Homogeneous photocatalytic degradation in the presence of dissolved metal ions, photosensitized reactions, or heterogeneous photocatalytic degradation on particulate metal compounds may contribute to the photochemical degradation of xenobiotics.

## 4.2. Examples of xenobiotic photodegradation studies

Pharmaceuticals and personal care products have been increasingly detected in aquatic field samples not only in Europe—e.g. in Italy [46, 47], Spain [48], United Kingdom [12, 49], Poland [50], and a EU-wide monitoring survey [16], but also in the USA [51, 52], China [53], and Japan [54].

The ecological impact of xenobiotics of the families of pharmaceuticals and personal care products is presently not sufficiently understood, partially because the environmental persistence of nearly all of these compounds has not yet been investigated. There are several indications that photochemical degradation may be a central factor in determining the environmental fate of these compounds. One of the supporting evidence lies in the structure of these substances—they often contain structural features that have been found photodegradable in other categories of compounds such as pesticides [30, 55].

As pointed out by Boreen et al. [24], the pesticides, carbaryl and napropamide, which photodegrade readily [55], contain the naphthoxy chromophore group that is found in pharmaceuticals, such as the non-steroid anti-inflammatory drugs, naproxen and nabumetone, and the beta-blocker, propranolol. The structure of carbaryl and naproxen is shown in **Figure 2**.

Because many of the pharmaceutical pollutants in surface waters have already eluded the biodegradation procedures of wastewater treatment, photochemistry in sunlit surface waters may be expected to play a much larger role than any naturally occurring biodegradation processes. Nevertheless, some compounds may evade photochemical degradation through sorption to suspended particles, which may be the case for substances such as the tetracyclines that have a high affinity for soil particles [56].

Many studies concerning emerging contaminants are focused on antibacterials, especially those used in both human and veterinary medicine, such as fluoroquinolones. Several representatives of fluoroquinolones are illustrated in **Figure 3**. Photochemical degradation of three representants of this group, norfloxacin, enrofloxacin, and ciprofloxacin, was studied, e.g., by

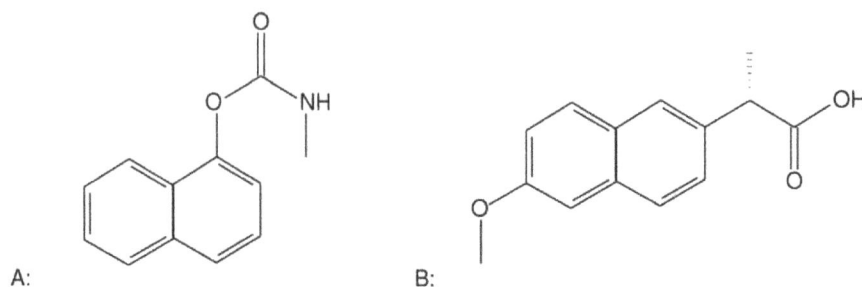

**Figure 2.** Chemical structure of the pesticide carbaryl (A) and anti-inflammatory drug naproxen (B).

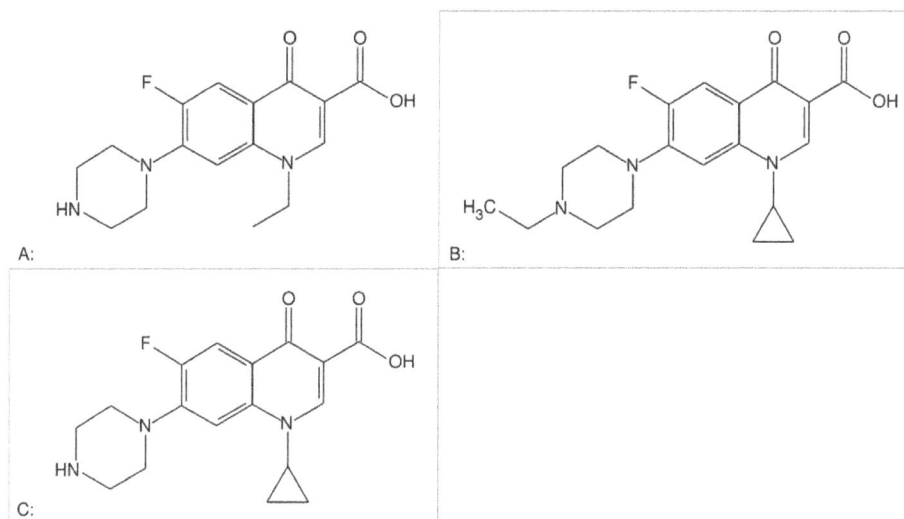

**Figure 3.** Examples of fluoroquinolone antibiotics: A—norfloxacin, B—enrofloxacin, and C—ciprofloxacin.

Babić et al. [57]. The source of radiation used in the study was a xenon lamp (emitting radiation in the 300–800 nm range); experiments were done in three matrices—MiliQ water, river water, and synthetic wastewater. It has been demonstrated that solar irradiation contributes significantly to the degradation of all three fluoroquinone derivatives—the mother compounds were degraded in 10 min after exposition to the radiation. Similar results were provided by the study of Sturini et al. [58] in which the reaction mixtures of two other fluoroquinolones, marbofloxacin and enrofloxacin, were investigated. The degradation was completed in about 1 h by exposure to solar light (Pavia, Italy, summer–noon time). The structure of the primary photoproducts showed that the degradation pathway proceeds via oxidative degradation of the piperazine side chain, reductive defluorination, and fluorine solvolysis.

Another important group of pharmaceutical products in connection with water pollution is antidepressants. Antidepressants are a class of pharmaceuticals used primarily to treat the symptoms of depression but can also be used to treat a wide variety of other medical conditions including sleep and eating disorders, alcohol and drug abuse, post-traumatic stress disorders, panic, and chronic pain. They are commonly prescribed for long-term use, leading to an increasing production volume compared to many other types of pharmaceuticals. According to Kessler et al. [59], almost 15 million American aged 18 and older suffer from a diagnosable major depressive disorder, thus giving rise to a market for branded antidepressants estimated to be worth US $14 billion [60].

Jeong-Wook Kwon and Armbrust [61] studied the laboratory persistence of fluoxetine (**Figure 4**), an antidepressant known under the brand names Prozac or Sarafen, which belongs to the selective serotonin reuptake inhibitor (SSRI) class of antidepressants. In the study, fluorescent lamps with a wavelength output of between 290 and 400 nm were adopted. In the experiments, fluoxetine was photochemically stable in buffered solutions as well as in two lake waters, the half-lives being greater than 100 days. This is not surprising since fluoxetine has a negligible absorption of radiation with wavelengths longer than 270 nm. The only exception was synthetic

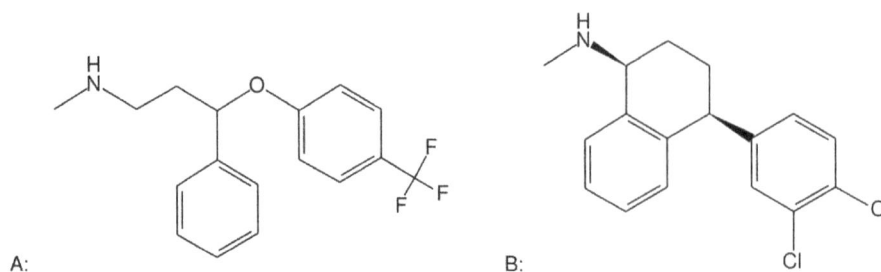

**Figure 4.** Structure of fluoxetine (A) and sertraline (B).

humic water in which the half-life was 21 day. Therefore, a photosensitised reaction with humic substances as sensitisers either for direct energy transfer or for reactive oxygen species (singlet oxygen) production can be hypothesised to be responsible for photoinitiated degradation in synthetic humic water.

The environmental fate of another SSRI antidepressant representant, sertraline, known under the brand name Zoloft, was explored by Jakimska et al. [62]. Simulated solar radiation (xenon lamp) was used for the experiments in eight different matrices: wastewater influent and effluent, untreated and treated water, river water, ultrapure water (pH 3 and 10), and methanol. The half-lives fell in the range of several days for most of the samples (from 4.9 days for wastewater effluent to 16.8 days for treated water); the only exceptions were ultrapure water with pH adjusted to 3 (127 days) and methanol (129 days). Since the authors observed a delay time in several cases, they proposed an autocatalytic mechanism as a plausible explanation for this observation.

Advanced oxidation treatment and the photochemical fate of three selected antidepressants in a solution of river humic acid was the subject of a study by Santoke et al. [63]. They focused on two antidepressants from the class of serotonin-norepinephrine reuptake inhibitors, SNRIs, duloxetine (brand name Cymbalta) and venlafaxine (brand name Effexor), which act on the two named neurotransmitters in the brain and are therefore more widely used than the older selective serotonin reuptake inhibitors, SSRIs, which act on only one neurotransmitter. The third substance, bupropion (brand name Wellbutrin or Zyban), is a norepinephrine-dopamine reuptake inhibitor, used both as an antidepressant and as a smoking cessation aid [64]. A Rayonet RPR-100 photochemical reactor with sixteen 350 nm fluorescent lamps and a solar simulator with a xenon lamp were employed for irradiation, commercially available Suwannee River humic acid was used as a sensitiser. In this study, concentrations of individual reactive species (singlet oxygen, hydroxyl radicals, hydrated electrons, and triplet excited state dissolved organic matter) were evaluated through specific probe reactions. Of the three antidepressant studied, only duloxetine was susceptible to direct photoreaction; venlafaxine and bupropion underwent indirect photoreaction to only a limited extent. The hydroxyl radicals were proven to be more important in the degradation of all three compounds in water to which humic acid had been added, compared to singlet oxygen or the hydrated electron. Pathways for the reaction of the antidepressants with hydroxyl radicals include hydroxylation and fragmentation. In the case of duloxetine, excited triplet state dissolved organic matter accounts partially for the photodegradation.

A significant group of organic xenobiotics in aquatic systems is parabens, p-hydroxybenzoic acid esters (**Figure 5**), widely used as preservatives in food products, cosmetics, toiletries, and pharmaceuticals. Parabens were first used as antimicrobial preservatives in pharmaceutical products in the mid-1920s and remained as preservative favourites for almost a century since they met several of the criteria of an ideal preservative: they exhibit a broad spectrum of antimicrobial activity, they have been considered safe to use, and they are stable over the broad pH range and sufficiently soluble in water to produce the effective concentration in the aqueous phase. In recent years, concern has been raised about their safety since several parabens have been reported to have estrogenic activity in experimental cell systems and animal models. Several studies, e.g. studies of Gomez et al. [65], Thuy et al. [66], and Chuang and Luo [67], investigated the photocatalytic degradation of parabens, namely ethylparaben and butylparaben, on $TiO_2$, focusing on operational parameters such as pH values and the initial concentration of parabens. Ethylparaben and butylparaben were demonstrated to have similar properties in terms of the values of adsorption constants and intrinsic reaction rates. The pH dependence was not significantly pronounced, but the reaction rate was slightly higher at pH = 4 than at other values (6, 9, 11). A study of transformation product led to a proposed pathway including an attack of the hydroxyl radical on the alkyl chain and the opening of the aromatic ring through hydroxylation to form alkyl carboxylic acid.

Klementova et al. [68] studied a set of pharmaceuticals of different classes including three parabens: methylparaben, ethylparaben, and propylparaben—both in homogeneous photocatalytic reaction mixture irradiated in the Rayonet RPR reactor with fluorescent lamps emitting wavelengths 300–350 nm and on $TiO_2$ with the lamps emitting in the region of 350–410 nm. Parabens were the most resistant substrates of all studied compounds. In the homogeneous reaction mixture, methylparaben exhibited mild photodegradation (40% of the substrate degraded in 90 min of irradiation) only in an extremely high, environmentally irrelevant, concentration of added Fe(III)—25 mg/l. Ethylparaben and propylparaben were more reactive than methylparaben—40% of ethylparaben and 60% of propylparaben were degraded after 90 min of irradiation in the presence of 5 mg of Fe(III) per 1 L of the reaction mixture. The measurement of the reduced form of iron (i.e. of the active catalytic form) in the reaction mixture revealed that steady state concentration of Fe(II) was attained in less than 5 min of irradiation; the steady state concentration of Fe(II) reached values between 60 and 70% of the total added ferric ions in the reaction mixtures of all parabens.

On $TiO_2$, methylparaben was again the least reactive substrate of the parabens studied—its degradation does not reach more than about 20% of the original amount in 120 min of

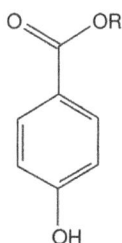

**Figure 5.** General chemical structure of a paraben.

irradiation. Ethylparaben and methylparaben reactivity was similar—48 and 52% of degraded ethylparaben and propylparaben, respectively.

An additional measurement of total organic carbon (TOC) in the reaction mixture [69] revealed that although the extent of the substrates degradation on $TiO_2$ is lower than in the homogeneous photocatalytic reaction, the decrease of organic carbon is higher in the reaction on $TiO_2$ compared with the homogeneous catalytic reaction. It means that mineralisation to $CO_2$ is more efficient with $TiO_2$ as the catalyst.

As accentuated earlier, organic xenobiotics reported in natural waters represent not only an extremely variegated, complex chemical system, including the above-mentioned drugs and preservatives, but also all groups of pesticides, chemicals used in the dye industry, explosives (TNT and its derivatives), solvents, and many others. For all these categories of compounds, the photochemical degradation processes, especially AOPs techniques, have been investigated. To mention just several of them, let us start with the photo-Fenton process. Yardin and Chiron [70] and Kröger and Fels [71] used it for the mineralisation of TNT and its derivatives. Haseneder et al. [72] and Santos et al. [73] employed the process for the degradation of polyethylene glycol, a substance with a wide range of application in both the industrial and pharmaceutical sectors. González et al. [74] and Dias et al. [75] achieved a significant mineralisation of the antibiotic sulfamethoxazole in wastewaters with this technique. Peternel et al. [35] and Sohrabi [76] conducted studies on the elimination of the persistent, non-biodegradable dyes, textile dye Red 45 and edible dye Carmoisine, respectively, by the photo-Fenton process.

A combination of ozone and UV radiation was shown to be effective in the degradation of dinitrotoluene and trinitrotoluene [77] as well as for some insecticides of the carbamate group such as carbofuran [78], for an industrial solvent N-methyl-2-pyrolidone [79], and for the reduction of trihalomethanes formation during drinking water treatment [80].

Heterogeneous photocatalysis with $TiO_2$ as the photocatalyst has been used for the degradation of sulfosalicylic acid in effluent [81]; the plant growth regulator 2,4-dichlorophenoxyacetic acid [82]; neonicotinoid insecticides [83]; the textile fibre reactive azo dye Procion Red MX-5B [84]; the extremely recalcitrant dye C.I. Reactive Red 2, RR2 [85]; and drugs such as paracetamol [86, 87]; tetracycline and beta-blockers [87], the calcium channel blocker verapamil, the corticosteroid cortisol, and the female sex hormone 17β-estradiol [68].

## 5. Conclusions

The photochemical reactions of pharmaceutical compounds as well as of many other contaminants are likely to play a major role in their fate in the aquatic environment. More information on their photodegradation pathways and on the degradation products and their persistence in the environment is essential for a better understanding of the impact of these contaminants on aquatic organisms and humans. The newly designed and quickly developing current advanced oxidation techniques are expected to help in the safe, efficient, and economic removal of the majority of these contaminants from wastewater effluents.

# Acknowledgements

The authors gratefully acknowledge the financial support of the research provided by Faculty of Science, University of South Bohemia.

# Conflict of interest

Hereby, I solemnly declare that I am the only author of the presented chapter and that no conflict of interest for a given manuscript exists that could inappropriately influence my judgement.

# Author details

Sarka Klementova

Address all correspondence to: sklement@jcu.cz

Faculty of Science, University of South Bohemie, Ceske Budejovice, Czech Republic

# References

[1] Palaniappan M, Gleick PH, Allen L, Cohen MJ, Christian-Smith J, Smith C. Clearing the Waters: A Focus on Water Quality Solutions. Report Prepared for the United Nations Environment Programme. Oakland, CA: Pacific Institute. Available from: http://pacinst.org/publication/clearing-the-waters-focus-on-water-quality-solutions/; 2010 [Accessed: December 2017]

[2] Managing Water under Uncertainty and Risk. In: World Water Development Report 4. Paris: UNESCO Publishing; UN, United Nations. 2012. Available from: http://unesdoc.unesco.org/images/0021/002156/215644e.pdf [Accessed: January 2018]

[3] Dudgeon D, Arthington AH, Gessner MO, Kawabata Z, Knowler DJ, Leveque C, Naiman RJ, Prieur-Richard AH, Soto D, Stiassny ML, Sullivan CA. Freshwater biodiversity: Importance, threats, status and conservation challenges. Biological Reviews of the Cambridge Philosophical Society. 2006;**81**:163-182

[4] Hughes SR. Occurrence and Effects of Pharmaceuticals in Freshwater Ecosystems [doctoral thesis]. University of Leeds. Available from: http://etheses.whiterose.ac.uk/5283/1/S%20R%20Hughes%20-%20Corrected%20PhD%20Thesis%20%28Oct%202013%29.pdf; 2013 [Accessed: December 2017]

[5] Water: A Matter of Life and Death. Fact Sheet. International Year of Freshwater. UN, United Nations. 2003. Available from: http://www.un.org/events/water/factsheet.pdf [Accessed: December 2017]

[6] Gavrilescu M, Demnerova K, Aamand J, Agathos S, Fava F. Emerging pollutants in the environment: Present and future challenges in biomonitoring, ecological risks and bioremediation. New Biotechnology. 2015;**32**:147-156. DOI: 10.1016/j.nbt.2014.01.001 [Accessed: December 2017]

[7] Milić N, Milanović M, Letić NJ, Sekulić MT, Radonić J, Mihajlović I, Miloradov MV. Occurrence of antibiotics as emerging contaminant substances in aquatic environment. International Journal of Environmental Health Research. 2013;**23**:296-310. DOI: 10.1080/09603123.2012.733934

[8] Kümmerer K. Emerging contaminants versus micro-pollutants. Clean Soil Air Water. 2011;**39**:889-890. Available from: http://onlinelibrary.wiley.com/doi/10.1002/clen.201110002/full [Accessed: December 2017]

[9] Marcoux MA, Matias M, Olivier F, Keck G. Review and prospect of emerging contaminants in waste—Key issues and challenges linked to their presence in wastewater treatment schemes: General aspects and focus on nanoparticles. Waste Management. 2013;**33**:2147-2156. DOI: 10.1016/j.wasman.2013.06.022

[10] US Geological Survey. Contaminants of emerging concern in ambient groundwater in urbanized areas of Minnesota, 2009-2012. 2014. Available from: https://pubs.usgs.gov/sir/2014/5096/pdf/sir2014-5096.pdf [ccessed December 2017)

[11] Tijany JO, Fatoba OO, Babajide OO, Petrik LF. Environmental Chemistry Letters. 2016;**14**:27-49. Available from: https://link.springer.com/article/10.1007/s10311-015-0537-z [Accessed: December 2017]

[12] Petrie B, Barden R, Kasprzyk/Hordern B. A review on emerging contaminants in wastewaters and the environment: Current knowledge, understanding areas and recommendations for future monitoring. Water Research. 2015;**72**:3-27. Available from: https://www.sciencedirect.com/science/article/pii/S0043135414006307 [[Accessed: January 2018]

[13] Kasprzyk-Hordern B, Baker DR. Enantiomeric profiling of chiral drugs in wastewater and receiving waters. Environmental Science & Technology. 2012;**43**:1681-1691

[14] Fernando-Climent L, Rodriguez-Mozaz S, Barceló D. Development of a UPLC-MS/MS method for the determination of ten anticancer drugs in hospital and urban wastewaters, and its application for the screening of human metabolites assisted by information-dependent acquisition tool (IDA) in sewage samples. Analytical and Bioanalytical Chemistry. 2013;**405**:5937-5952

[15] Fenech C, Nolan K, Rock L, Morrissey A. An SPE LC-MA/MS method for the analysis of human and veterinary chemical markers within surface waters: An environmental forensics application. Environmental Pollution. 2013;**181**:250-256

[16] Loos R, Carvalho R, António DC, Comero S, Locoro G, Tavazzi S, Paracchini B, Gieani M, Lettieri T, Blaha L, Jarosova B, Voorspoels S, Servaes K, Haglund PL, Fick J, Lindberg RH, Schweig D, Gawlik BM. EU-wide monitoring survey on emerging polar organic contaminants in wastewater treatment plant effluents. Water Research. 2013;**47**:6475-6487. DOI: 10.1016/j.watres.2013.08.024

[17] López-Serna R, Petrovic M, Barceló D. Development of a fast instrumental method for the analysis of pharmaceuticals in environmental and wastewaters based on ultrahigh performance liquid chromatography (UHPLC)-tandem mass spectrometry (MS/MS). Chemosphere. 2011;**85**:1390-1399

[18] Phillips PJ, Smith SG, Kolpin DW, Zaug SD, Buxton HT, Furlong ET, et al. Pharmaceutical formulation facilities as sources of opioids and other pharmaceuticals to wastewater treatment plant effluents. Environmental Science & Technology. 2010;**44**:4910-4916

[19] Li D, Yang M, Hu J, Ren L, Zhang Y, Li K. Determination and fate of oxytetracycline and related compounds in oxytetracycline production wastewater and the receiving river. Environmental Toxicology and Chemistry. 2008;**27**(1):80-86

[20] Aly OM, Faust SD. Herbicides in surface waters; studies on fate of 2,4-D and ester derivatives in natural surface waters. Journal of Agricultural and Food Chemistry. 1964; **12**:541-546

[21] Wauchope RD. The pesticide content of surface water draining from agricultural fields— A review. Journal of Environmental Quality. 1978;**7**:459-472

[22] Thurman EM, Goolsby DA, Meyer MT, Kolpin DW. Herbicides in surface waters of the midwestern United States: The effect of spring flush. Environmental Science & Technology. 1991;**25**:1794-1796

[23] Readman JW, Albanis TA, Barcelo D, Galassi S, Tronczynski J, Gabrielides GP. Herbicide contamination of Mediterranean estuarine waters: Results from MED. POL. Pilot survey. Marine Pollution Bulletin. 1993;**26**:613-619

[24] Boreen AL, Arnold WA, McNeill K. Photodegradation of pharmaceuticals in the aquatic environment: A review. Aquatic Sciences. 2003;**65**:320-341. Available from: https://link. springer.com/content/pdf/10.1007%2Fs00027-003-0672-7.pdf [Accessed: December 2017]

[25] Klementova S, Zlamal M. Photochemical degradation of triazine herbicides—Comparison of homogeneous and heterogeneous photocatalysis. Photochemical & Photobiological Sciences. 2013;**12**:660-663

[26] Stevenson FJ. Humus Chemistry. New York: Wiley; 1982

[27] Perminova IV, Frimmel FH, Kudryavtsev AV, Kulikova NA, Abbt-Braun G, Hesse S, Petrosyan VS. Molecular weight characteristics of humic substances from different environments as determined by size exclusion chromatography and their statistical evaluation. Environmental Science & Technology. **37**:2477-2485

[28] Ghabbour EA, Davies G, editors. Humic Substances: Structures, Models and Functions. Cambridge, U.K.: RSC Publishing; 2001

[29] Klementova S, Hamsova K. Catalysis and sensitization in photochemical degradation of Triazines. Research Journal of Chemistry and Environment. 2000;**4**:7-12

[30] Klementova S. A critical view of the photoinitiated degradation of herbicides. In: Mohammed Naguib Abd El-Ghany Hasaneen, editor. Herbicides—Properties, Synthesis and

Control of Weeds. 2012. ISBN: 978-953-307-803-8, InTech. Available from: http://www. intechopen.com/articles/show/title/a-critical-view-of-the-photoinitiated-degradation-of-herbicides [Accessed: December 2017]

[31] Hashimoto K, Irie H, Fujishima A. TiO$_2$ Photocatalysis: A historical overview and future prospects. Japanese Journal of Applied Physics. 2005;**44**:8269-8285. Available from: https://www.jsap.or.jp/jsapi/Pdf/Number14/04_JJAP-IRP.pdf [Accessed: 19 December 2017]

[32] Chen J, Poon C-s. Photocatalytic construction and building materials: From fundamentals to applications. Building and Environment. **44**:1899-1906. Available from: http://www.sciencedirect.com/science/article/pii/S0360132309000134 [Accessed: December 2017]

[33] Lacombe S, Keller N. Photocatalysis: Fundamentals and applications in JEP 2011. Environmental Science and Pollution Research. 2012;**19**:3651-3654. Available from: https://www.researchgate.net/publication/229160470_Photocatalysis_Fundamentals_and_applications_in_JEP_2011 [Accessed: December 2017]

[34] Chakrabarti S, Dutta BK. Photocatalytic degradation of model textile dyes in wastewater using ZnO as semiconductor catalyst. Journal of Hazardous Materials. 2004;**112**:269-278. Available from: http://www.sciencedirect.com/science/article/pii/S0304389404002584 [Accessed: December 2017]

[35] Peternel IT, Koprivanac N, Bozić AM, Kusić HM. Comparative study of UV/TiO$_2$, UV/ZnO and photo-Fenton processes for the organic reactive dye degradation in aqueous solution. Journal of Hazardous Materials. 2007;**148**:477-484. DOI: 10.1016/j.jhazmat.2007.02.072. PMID 17400374

[36] Hu J-S, Ren L-L, Guo Y-G, Liang H-P, Cao A-M, Wan L-J, Bai C-L. Mass production and high photocatalytic activity of ZnS nanoporous nanoparticles. Angewandte Chemie. 2005;**117**:1295-1299. Available from: http://onlinelibrary.wiley.com/doi/10.1002/ange.200462057/full [Accessed: December 2017]

[37] Hu Y-S, Kleiman-Shwarsctein A, Forman AJ, Hazen D, Park J-N, McFarland EW. Pt-doper $\alpha$-Fe$_2$O$_3$ thin films active for photoelectrochemical water splitting. Chemistry of Materials. 2008;**20**:3803-3805

[38] Byrappa K, Subramani AK, Ananda S, Lokanatha Rai KM, Dinesh R, Yoshimura M. Photocatalytic degradation of Rhodamine B dye using hydrothermally synthesized ZnO. Bulletin of Material. Science. 2006;**29**:433-438. Available from: http://www.ias.ac.in/matersci/bmsoct2006/433.pdf [Accessed: December 2017]

[39] Zha Y, Zhang S, Pang H. Preparation, characterization and photocatalytic activity of CeO$_2$ nanocrystalline using ammonium bicarbonate as precipitant. Material Letters. 2007;**61**:1863-1866. Available from: http://www.sciencedirect.com/science/article/pii/S0167577X06009633 [Accessed: December 2017]

[40] Guo Y, Quan X, Lu N, Zhao H, Chen S. High photocatalytic capability of self-assembled nanoporous WO$_3$ with preferential orientation of (002) planes. Environmental Science and Technology. 2007;**41**:4422-4427

[41] Granados-Oliveros G, Páez-Mozo EA, Martínez-Ortega F, Ferronato C, Chovelon JM. Degradation of atrazine using metalloporphyrins supported on $TiO_2$ under visible light irradiation. Applied Catalysis B: Environmental. 2009;**89**:448-454

[42] Moreno-Escobar B, Gomez Nieto MA, Hontoria Garcia E. Simple tertiary treatment systems. Water Science and Technology. Water Supply. 2005;**5**:35-41. Available from: http://ws.iwaponline.com/content/5/3-4/35 [Accessed: December 2017]

[43] Walid KL, Al-Quodah Z. Combined advanced oxidation and biological treatment processes for the removal of pesticides from aqueous solutions. Journal Hazardous Material. 2006;**137**:489-497. DOI: 10.1016/j.jhazmat.2006.02.027

[44] Poyatos JM, Munio MM, Almecija MC, Torres JC, Hotoria E, Osorio F. Advanced oxidation processes for wastewater treatment: State of the art. Water, Air, and Soil Pollution. 2010;**205**:187-204

[45] Canada J, Pedros G, Bosca JV. Relationships between UV (0.290-0.385 μm) and broad band solar radiation hourly values in Valencia and Códoba, Spain. Energy. 2003;**28**:199-217. Available from: https://ac.els-cdn.com/S0360544202001111/1-s2.0-S0360544202001111-main. pdf?_tid=7bea51c6-e1a2-11e7-bdb9-00000aacb35d&acdnat=1513347544_a05da2847b6837a8 bf8c07251766eaf6 [Accessed: December 2017]

[46] Zuccato E, Calamari D, Natangelo M, Fanelli R. Presence of therapeutic drugs in the environment. Lancet. 2000;**355**:1789-1790

[47] Calamari D, Zuccato E, Castiglioni S, Bagnati R, Fanelli R. Strategic survey of therapeutic drugs in the rivers Po and Lambro in northern Italy. Environmental Science & Technology. 2003;**37**:1241-1248

[48] Albero B, Pérez RA, Sánchez-Brunete C, Tadeo JL. Occurrence and analysis of parabens in municipal sewage sludge from wastewater treatment plants in Madrid (Spain). Journal of Hazardous Materials. 2012;**239**:48-55. Available from: http://www.sciencedirect.com/science/article/pii/S0304389412004992 [Accessed: December 2017]

[49] Baker DR, Barron L, Kasprzyk-Hordern B. Illicit and pharmaceutical drug consumption estimated via wastewater analysis. Part A: Chemical analysis and drug use estimates. Science of the Total Environment. 2014;**487**:629-641. DOI: 10.1016/j.scitotenv.2013.11.107

[50] Zgola-Grześkowiak A, Jeszka-Skowron M, Czazczyńska-Goślińska B, Grześkowiak T. Determination of parabens in Polish River and lake water as a function of season. Analytical Letters. 2016;**49**:1734-1747

[51] Loraine GA, Pettigrove ME. Seasonal variations in concentrations of pharmaceuticals and personal care products in drinking water and reclaimed wastewater in Southern California. Environmental Science & Technology. 2006;**40**:687-695

[52] Fram MS, Belitz K. Occurrence and concentrations of pharmaceutical compounds in groundwater used for public drinking supply in California. Science of the Total Environment. 2011;**409**:3409-3417

[53] Peng X, Yu Y, Tang C, Tan J, Huang Q, Wang Z. Occurrence of steroid estrogens,

endocrine-disrupting phenols, and acid pharmaceuticals in Uran riverine water of the Pearl river Delta, South China. Science of the Total Environment. 2008;**397**:158-166. Available from: https://ac.els-cdn.com/S0048969708002441/1-s2.0-S0048969708002441-main.pdf?_tid=8eb936e6-ebb6-11e7-a31e-00000aab0f6b&acdnat=1514455677_86218ac05be58a0ae99b43 21c1d8fe62 [Accessed: December 2017]

[54] Tamura I, Yasuda Y, Kagota KI, Yoneda S, Nakada N, Kumar V, Kameda Y, Kimura K, Tatarazako N, Yamamoto H. Contribution of pharmaceuticals and personal care products (PPCPs) to whole toxicity of water samples collected in effluent-dominated urban streams. Ecotoxicology and Environmental Safety. 2017;**144**:338-350. DOI: 10.1016/j.ecoenv.2017.06.032

[55] Burrows HD, Canle LM, Santaballa JA, Seenken S. Reaction pathways and mechanisms of photodegradation of pesticides. Journal of Photochemistry and Photobiology B: Biology. 2002;**67**:71-108

[56] Tolls J. Sorption of veterinary pharmaceuticals in soils: A review. Environmental Science & Technology. 2001;**35**:3397-3406

[57] Babić S, Periša M, Škorić I. Photolytic degradation of norfloxacin, enrofloxacin and cipro-floxacin in various aqueous media. Chemosphere. 2013;**91**:1635-1642

[58] Sturini M, Speltini A, Maraschi F, Profumo A, Pretali L, Fasani E, Albini A. Photochemical degradation of Marbofloxacin and Enrofloxacin in natural waters. Environmental Science & Technology. 2010;**44**:4564-4569

[59] Kesler RC, Chiu Wai T, Demler O, Meriknagas KR, Walters EE. Prevalence, severity, and comorbidity of 12-month DSM-IV disorders in the National Comorbidity Survey Replication. Archives of General Psychiatry. 2005;**62**:617-627

[60] Bartholow M. Top 200 Prescription Drugs of 2009. Pharmacy Times—Practical Information for Today's Pharmacist. 2010. Available from: http://www.pharmacytimes.com/publications/issue/2010/may2010/rxfocustopdrugs-0510 [Accessed: December 2017]

[61] Kwon J-W, Armbrust KL. Laboratory persistence and fate of fluoxetine in aquatic environments. Environmental Toxicology and Chemistry. 2006;**25**:2561-2568. Available from: http://onlinelibrary.wiley.com/doi/10.1897/05-613R.1/full [Accessed: December 2017]

[62] Jakimska A, Śliwka Kaszyńska M, Nagórski P, Kot Wasik A, Nmeiśnik J. Environmental Fate of Two Psychiatric Drugs, Diazepam and Sertraline: Phototransformation and Investigation of their Photoproducts in Natural Waters. J. Chromatogr. Sep. Tech. 2014;**5** (no pages, open access). DOI:10.4172/2157-7064.1000253

[63] Santoke H, Weihua S, Cooper W J, Peake BM. Advanced oxidation treatment and photo-chemical fate of selected antidepressant pharmaceuticals in solutions of Suwannee River humic acid. Journal of Hazardous Materials. 2012;**217-218**:382-390

[64] Gonzales D, Rennard SI, Nides M, Oncken C Azoulay S, Billing CB, Watsky EJ, Gong J, Williams KE, Reeves KR. Varenicline, an alpha-4-beta-2 nicotinic acetylcholine receptor partial agonist, vs sustained-release bupropion and placebo for smoking cessation. The Journal of the American Medical Association. 2006;**296**:47-55. Available

from: https://jamanetwork.com/journals/jama/fullarticle/211000 [Accessed: December 2017]

[65] Gomez E, Pillon A, Fenet H, Rosain D, Duchesne MJ, Nicolas JC, Balaguer P, Casellas C. Estrogenic activity of cosmetic components in reporter cell lines: Parabens, UV screens, and Musks. Journal of Toxicology and Environmental Health, Part A. 2005;**68**:239-251

[66] Vo TTB, Yoo Y-M, Choi K-C, Jeung E-B. Potential estrogenic effect(s) of parabens at the prepubertal stage of a postnatal female rat model. Reproductive Toxicology. 2010;**29**:306-316. DOI: 10.1016/j.reprotox.2010.01.013

[67] Chuang LC, Luo CH. Photocatalytic degradation of parabens in aquatic environment: Kinetics and degradation pathway. Kinetics and Catalysis. 2015;**56**:412-418. Available from: https://link.springer.com/content/pdf/10.1134%2FS0023158415040047.pdf [Accessed: December 2017]

[68] Klementova S, Kahoun D, Doubkova L, Frejlachova K, Dusakova M, Zlamal M. Catalytic photodegradation of pharmaceuticals—Homogeneous and heterogeneous photocatalysis. Photochemical & Photobiological Sciences. 2017;**16**:67-71

[69] Frejlachová K. Photochemical degradation of parabens [Mgr Thesis] (in Czech). České Budějovice, Czech Rep: Faculty of Science, University of South Bohemia; 2017

[70] Yardin G, Chiron S. Photo-Fenton treatment of TNT contaminated soil extract solutions obtained by soil flushing with cyclodextrin. Chemistry. 2006;**62**:1395-1402. DOI: 10.1016/j.chemosphere.2005.05.019

[71] Kröger M, Fels G. Combined biological—Chemical procedure for the mineralization of TNT. Biodegradation. 2007;**18**:413-425

[72] Haseneder R, Fdez-Navamuel B, Härtel G. Degradation of polyethylene glycol by Fenton reaction: A comparative study. Water Science and Technology. 2007;**55**:83-87

[73] Santos LC, Schmitt CC, Poli AL, Neumann MG. Photo-fenton degradation of poly (ethyleneglycol). The Journal of the Brazilian Chemical Society. 2011;**22**:no pages, open access. DOI: 10.1590/S0103-50532011000300018

[74] González O, Sans C, Espulgas S. Sulfamethoxazole abatement by photo-Fenton: Toxicity, inhibition and biodegradability assessment of intermediates. Journal of Hazardous Materials. 2007;**146**:459-464

[75] Dias IN, Souza BS, Pereira JHOS, Mreira FC, Dezotti M, Boaventura RAR, Vilar VJP. Enhancement of the photo-Fenton reaction at near neutral pH through the use of ferrioxalate complexes: A case study on trimethoprim and sulfamethoxazole antibiotics removal from aqueous solution. Chemical Engineering Journal. 2014;**247**:302-313. DOI: 10.1016/j.cej.2014.03.020

[76] Sohrabi MR, Shariati Sb KA, Sn S. Removal of Carmoisine edible dye by Fenton and photo Fenton processes using Taguchi orthogonal array design. Arabian Journal of Chemistry. 2017;**10**:S3523-S3531. DOI: 10.1016/j.arabjc.2014.02.019

[77] Chen W, Juan C, Wei K. Decomposition of dinitrotoluene isomers and 2,4,6-trinitrotoluene in spent acid from toluene nitration process by ozonation and photo-ozonation. Journal of Hazardous Materials. 2007;**147**:97-104

[78] Lau T, Graham N. Degradation of the endocrine disruptor carbofuran by UV, $O_3$ and $O_3$/UV. Water Science and Technology. 2007;**55**:275-280

[79] Wu JJ, Muruganandham M, Chang LT, Yang JS, Chen SH. Ozone-based advanced oxidation processes for the decomposition of N-methyl-2-pyrolidone in aqueous medium. Ozone: Science & Engineering. 2007;**29**:177-183

[80] Borikar D, Mohseni M, Jasim S. Evaluations of conventional, ozone and UV/$H_2O_2$ for removal of emerging contaminants and THM-FPs. Water Quality Research Journal. 2015;**50**:140-151

[81] Tong SP, Xie DM, Wei H, Liu WP. Degradation of sulfosalicylic effluents by $O_3$/UV, $O_3$/$TiO_2$/UV and $O_3$/V-O/$TiO_2$: A comparative study. Ozone Science & Engineering. 2005;**27**:233-238

[82] Giri RR, Ozaki H, Ishida T, Takanami R, Taniguchi S. Synergy ozonation and photocatalysis to mineralize low concentration 2,4-dichlorophenoxiacetic acid in aqueous solution. Chemosphere. 2007;**66**:1610-1617

[83] Černigoj U, Štangar UL, Trebše P. Degradation of neonicotinoid insecticides by different advanced oxidation processes and studying the effect of ozone on $TiO_2$ photocatalysis. Applied Catalysis B Environmental. 2007;**75**:229-238

[84] Lin YC, Lee HS. Effects of $TiO_2$ coating dosage and operational parameters on a $TiO_2$/Ag photocatalysis system for decolorizing Procion red MS-5B. Journal of Hazardous Materials. 2010;**179**:462-470. DOI: 10.1016/j.jhazmat.2010.03.026

[85] Wang X, Jia J, Wang Y. Degradation of C.I. Reactive red 2 through photocatalysis coupled with water jet cavitation. Journal of Hazardous Materials. 2011;**185**:315-321. DOI: 10.1016/j.jhazmat.2010.09.036

[86] Borges ME, García DM, Hernández T, Ruiz-Morales JC, Esparza P. Supported photocatalyst for removal of emerging contaminants from wastewater in a continuous packed-bed photoreactor configuration. Catalysts. 2015;**5**:77-87. Available from: https://pdfs.semanticscholar.org/b9c2/0c51c20dc10fecc4fee060ed17658ed7ad8a.pdf [Accessed: December 2017]

[87] Rimoldi L, Meroni D, Falletta E, Pifferi V, Falciola L, Cappelletti G, Ardizzone S. Emerging pollutant mixture mineralization by $TiO_2$ photocatalysts. The role of the water medium. Photochemical & Photobiological Sciences. 2017;**16**:60-66. Available from: http://pubs.rsc.org/-/content/articlehtml/2017/pp/c6pp00214e [Accessed: 29 December 2017]

# Permissions

The contributors of this book come from diverse backgrounds, making this book a truly international effort. This book will bring forth new frontiers with its revolutionizing research information and detailed analysis of the nascent developments around the world.

We would like to thank all the contributing authors for lending their expertise to make the book truly unique. They have played a crucial role in the development of this book. Without their invaluable contributions this book wouldn't have been possible. They have made vital efforts to compile up to date information on the varied aspects of this subject to make this book a valuable addition to the collection of many professionals and students.

This book was conceptualized with the vision of imparting up-to-date information and advanced data in this field. To ensure the same, a matchless editorial board was set up. Every individual on the board went through rigorous rounds of assessment to prove their worth. After which they invested a large part of their time researching and compiling the most relevant data for our readers.

The editorial board has been involved in producing this book since its inception. They have spent rigorous hours researching and exploring the diverse topics which have resulted in the successful publishing of this book. They have passed on their knowledge of decades through this book. To expedite this challenging task, the publisher supported the team at every step. A small team of assistant editors was also appointed to further simplify the editing procedure and attain best results for the readers.

Apart from the editorial board, the designing team has also invested a significant amount of their time in understanding the subject and creating the most relevant covers. They scrutinized every image to scout for the most suitable representation of the subject and create an appropriate cover for the book.

The publishing team has been an ardent support to the editorial, designing and production team. Their endless efforts to recruit the best for this project, has resulted in the accomplishment of this book. They are a veteran in the field of academics and their pool of knowledge is as vast as their experience in printing. Their expertise and guidance has proved useful at every step. Their uncompromising quality standards have made this book an exceptional effort. Their encouragement from time to time has been an inspiration for everyone.

The publisher and the editorial board hope that this book will prove to be a valuable piece of knowledge for researchers, students, practitioners and scholars across the globe.

# List of Contributors

Sankalan Mondal and Satyen Saha
Department of Chemistry, Institute of Science, Banaras Hindu University, Varanasi, India

Mallory G. John, Victoria Kathryn Meader and Katharine Moore Tibbetts
Department of Chemistry, Virginia Commonwealth University, Richmond, VA, USA

Alexandrina Nuta and Ana-Alexandra Sorescu
ICECHIM, Evaluation and Conservation of Cultural Heritage, Bucharest, Romania

Rodica-Mariana Ion and Lorena Iancu
ICECHIM, Evaluation and Conservation of Cultural Heritage, Bucharest, Romania
Materials Engineering Doctoral School, Valahia University, Targoviste, Romania

Issah Yahaya and Zeynel Seferoglu
Department of Chemistry, Gazi University, Ankara, Turkey

Muhammad Saeed, Muhammad Usman and Atta ul Haq
Department of Chemistry, Government College University Faisalabad, Faisalabad, Pakistan

Rebeca Sola Llano, Edurne Avellanal Zaballa, Jorge Bañuelos and Iñigo López Arbeloa
Department of Physical Chemistry, University of the Basque Country (UPV/EHU), Bilbao, Spain

César Fernando Azael Gómez Durán, José Luis Belmonte Vázquez and Eduardo Peña Cabrera
Department of Chemistry, University of Guanajuato, Guanajuato, Mexico

Sebok Lee, Myungsam Jen, Kooknam Jeon, Joonwoo Kim and Yoonsoo Pang
Department of Chemistry, Gwangju Institute of Science and Technology, Gwangju, Republic of Korea

Jaebeom Lee
Department of Physics and Photon Science, Gwangju Institute of Science and Technology, Gwangju, Republic of Korea

Nadezhda Mikhailovna Storozhok and Nadezhda Medyanik
Tyumen State Medical University, Ministry of Health of the Russian Federation, Odesskaya, Tyumen, Russian Federation

Ming-Der Su
Department of Applied Chemistry, National Chiayi University, Chiayi, Taiwan
Department of Medicinal and Applied Chemistry, Kaohsiung Medical University, Kaohsiung, Taiwan

Carolina Santacruz-Perez, Paulo Newton Tonolli, Felipe Gustavo Ravagnani and Maurício S. Baptista
Biochemistry Department, Institute of Chemistry, University of São Paulo, São Paulo, Brazil

Sarka Klementova
Faculty of Science, University of South Bohemie, Ceske Budejovice, Czech Republic

# Index